T0257858

Telecommunication Networks:
Present and Future Scenario

Telecommunication Networks: Present and Future Scenario

Edited by **Bernhard Ekman**

CLANRYE
INTERNATIONAL

New Jersey

Published by Clanrye International,
55 Van Reypen Street,
Jersey City, NJ 07306, USA
www.clanryeinternational.com

Telecommunication Networks: Present and Future Scenario
Edited by Bernhard Ekman

International Standard Book Number: 978-1-63240-483-1 (Hardback)

Printed in the United States of America.

Contents

Preface

A descriptive account on telecommunications networks has been presented in this profound book. It elucidates the basics of fast developing networks as well as the advanced concepts and future expectations of Telecommunications Networks. It recognizes and analyzes the most important research issues in Telecommunication and it consists of information contributed by the top researchers, industry and academic professionals. This book also contains surveys of current publications that thoroughly examine important fields of interest like: 3G/4G, IMS, modeling, quality of service, eTOM etc. The book covers topics like New Generation Networks, Quality of Services, and Sensor Networks. It will serve as a good reference for both PhD and MA students.

This book is the end result of constructive efforts and intensive research done by experts in this field. The aim of this book is to enlighten the readers with recent information in this area of research. The information provided in this profound book would serve as a valuable reference to students and researchers in this field.

At the end, I would like to thank all the authors for devoting their precious time and providing their valuable contribution to this book. I would also like to express my gratitude to my fellow colleagues who encouraged me throughout the process.

Editor

Part 1

New Generation Networks

Access Control Solutions for Next Generation Networks

F. Pereniguez-Garcia, R. Marin-Lopez and A.F. Gomez-Skarmeta
Faculty of Computer Science, University of Murcia
Spain

1. Introduction

In recent years, wireless telecommunications systems have been prevalently motivated by the proliferation of a wide variety of wireless technologies, which use the air as a propagation medium. Additionally, users have been greatly attracted for wireless-based communications since they offer an improved user experience where information can be exchanged while changing the point of connection to the network. This increasing interest has led to the appearance of mobile devices such as smart phones, tablet PCs or netbooks which, equipped with multiple interfaces, allow *mobile users* to access network services and exchange information anywhere and at any time. To support this *always-connected* experience, communications networks are moving towards an *all-IP* scheme where an IP-based network core will act as connection point for a set of accessible networks based on different wireless technologies. This future scenario, referred to as the *Next Generation Networks* (NGNs), enables the convergence of different heterogeneous wireless access networks that combine all the advantages offered by each wireless access technology per se.

In a typical NGN scenario users are expected to be potentially mobile. Equipped with wireless-based multi-interface lightweight devices, users will go about their daily life (which implies to perform movements and changes of location) while demanding access to network services such as VoIP or video streaming. The concept of *mobility* demands session continuity when the user is moving across different networks. In other words, active communications need to be maintained without disruption (or limited breakdown) when the user changes its connection point to the network during the so-called *handoff*.

This aspect is of vital importance in the context of NGNs to allow the user to roam seamlessly between different networks without experiencing temporal interruption or significant delays in active communications. Nevertheless, during the handoff, the connection to the network may for various reasons be interrupted, which causes a packet loss that finally impacts on the on-going communications.

Thus, to achieve mobility without interruptions and improve the quality of the service perceived by the user, it is crucial to reduce the time required to complete the handoff. The handoff process requires the execution of several tasks (N. Nasser et al. (2006)) that negatively affect the handoff latency. In particular, the authentication and key distribution processes have been proven to be one of the most critical components since they require considerable time (A. Dutta et al. (2008); Badra et al. (2007); C. Politis et al. (2004); Marin-Lopez et al. (2010); R. M. Lopez et al. (2007)). The implantation of these processes during the *network access control* demanded by network operators is destined to ensure that only allowed users can access the network resources in a secure manner. Thus, while necessary, these security services must be carefully taken into account, since they may significantly affect the achievement of seamless mobility in NGNs.

In this chapter we are going to revise the different approaches that have been proposed to address this challenging issue in future NGNs. More precisely, we are going to carry out this analysis in the context of the *Extensible Authentication Protocol* (EAP), a protocol which is acquiring an important position for implementing the access control solution in future NGNs. This interest is motivated by the important features offered by the protocol such as flexibility and media independence. Nevertheless, the EAP authentication process has shown certain inefficiency in mobile scenarios. In particular, a typical EAP authentication involves a considerable signalling to be completed. The research community has addressed this problem by defining the so-called *fast re-authentication* solutions aimed at reducing the latency introduced by the EAP authentication. Throughout this chapter, we will revise the different groups of fast re-authentication solutions according to the strategy followed to minimize the authentication time.

The remaining of the chapter is organized as follows. Section 2 describes the different technologies related to the network access authentication. Next, Section 3 outlines the deficiencies of EAP in mobile environments, which have motivated the research community the proposal of fast re-authentication solutions. The different fast re-authentication schemes proposed so far are analyzed in Section 4. Finally, the chapter finalizes with Section 5 where the most relevant conclusions are extracted.

2. Protocols involved in the network access service

2.1 AAA infrastructures: Authentication, Authorization and Accounting (AAA)

Network operators need to control their subscribers so that only authenticated and authorized ones can access to the network services. Typically, the correct support of a controlled access to the network service has been guaranteed by the deployment of the so-called *Authentication, Authorization and Accounting* (AAA) infrastructures (C. de Laat et al. (2000)). AAA essentially defines a framework for coordinating these individual security services across multiple network technologies and platforms.

An overview of the different components is the best way to understand the services provided by the AAA framework.

- *Authentication.* This service provides a means of identifying a user that requires access to some service (e.g., network access). During the authentication process, users provide a set of credentials (e.g., password or certificates) in order to verify they are who they claim to be. Only when the credentials are correctly verified by the AAA server, the user is granted access to the service.

- *Authorization.* Authorization typically follows the authentication and entails the process of determining whether the client is allowed to perform and request certain tasks or operations. Authorization is the process of enforcing policies, determining what types or qualities of activities, resources or services a user is permitted.

- *Accounting.* The third component in the AAA framework is accounting, which measures the resources a user consumes during network access. This can include the amount of time a service is used or the amount of data a user has sent and/or received during a session. Accounting is carried out by gathering session statistics and usage information, and it is used for different purposes like billing.

The following sections provide a detailed description for the general AAA architecture and the most relevant AAA protocols.

2.1.1 Generic AAA architecture

The general AAA scheme, as defined in (C. de Laat et al. (2000)), requires the participation of four different entities (see Fig. 1) that take part in the authentication, authorization and accounting processes:

- A *user* desiring to access a specific service offered by the network operator.
- A *domain* where the user is registered. This domain, typically referred to as *home domain*, is able to verify the user's identity based on some credentials. Optionally, the home domain not only authenticates but also provides authorization information to the user
- A *service provider* controlling the access to the offered services. The service provider can be implemented by the domain where the user is subscribed to (home domain) or by a different domain in the roaming cases. In the case the service provider is located outside the home domain, the access to the service is provided on condition that an agreement is established between the service provider and the home domain. These bilateral agreements, which may take the form of formal contracts known as *Service Level Agreements* (SLAs), suppose the establishment of a trust relationship between the involved domains that will allow the service provider to authenticate and authorize foreign users coming from another administrative domains.
- A *service provider's service equipment* which will be typically located on a device that belongs to the service provider. For example, in the case of network access service, this role is played by the *Network Access Server* (NAS) like, for example, an 802.11 access point.

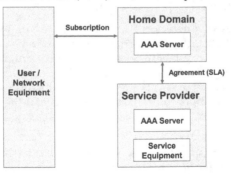

Fig. 1. Generic AAA architecture

2.1.2 Relevant AAA protocols

To allow the communication between AAA servers, it is required the deployment of a *AAA protocol*. Nowadays, the most relevant AAA protocols are RADIUS (C. Rigney et al. (2000)) and Diameter (P. Calhoun & J. Loughney (2003)). Despite Diameter is the most complete AAA protocol, RADIUS is the most widely deployed one in current AAA infrastructures. In the following, it is provided a brief overview of both.

2.1.2.1 RADIUS

RADIUS is a client-server protocol where a NAS usually acts as *RADIUS client*. During authentication procedures, the RADIUS client is responsible for passing user information in the form of requests to the *RADIUS server* and waits for a response from the server. Depending on the policy, the NAS may only need a successful authentication or further authorization directives from the server to enable data traffic to the client. The RADIUS server, on the other hand, is responsible for processing requests, authenticating the users and returning the information necessary for user-specific configuration to deliver the service.

The typical RADIUS conversation consists of the following messages:

- *Access-Request*. This message is sent from the RADIUS client (NAS) to the server to request authentication and authorization for a particular user.

- *Access-Challenge*. This message, sent from the RADIUS server to the client, is used by the server to obtain more information from the NAS about the end user in order to make a decision about the requested service.

- *Access-Accept*. This message is sent from the RADIUS server to the NAS to indicate a successful completion of the request.

- *Access-Reject*. This message is sent by the server to indicate the rejection of a request.

Typically, the main part of a RADIUS conversation consists of several Access-Request/Access-Challenge message exchanges where the RADIUS client and server exchange information transported within RADIUS attributes. Depending on whether the client is successfully authenticated or not, the RADIUS server finalizes the communication with an *Access-Accept* or *Access-Reject*, respectively.

Apart from these main messages, the RADIUS base specification defines some others to transmit accounting information (*Accounting-Request*/*Accounting-Response*) or the status of the RADIUS entities (*Status-Client*/*Status-Server*).

Regarding the protocol used to transport RADIUS messages, protocol designers considered that the *User Datagram Protocol* (UDP) was the most appropriate one since the *Transmission Control Protocol* (TCP) session establishment is a time-consuming process requiring the management of connection state. Nevertheless, the lack of a reliable transport causes serious problems to RADIUS. For example, clients are unable to distinguish when a request is received by the server or a communication problem has occurred and the RADIUS packet has not reached its destination. Similarly, a client cannot distinguish whether a server is down or discarding requests.

RADIUS security is another aspect that was not deeply considered. In particular, it is based on the use of shared secrets between the RADIUS client and the server. In real deployments, this basic security mechanism has been known to cause several vulnerabilities:

- Shared secrets must be statically configured. No method for dynamic shared secret establishment is defined in the RADIUS protocol.

- Shared secrets are determined according to the source IP address in the RADIUS packet. This introduces management problems when the client's IP address change.

- When using RADIUS proxies, the RADIUS client only shares a secret with the RADIUS server in the first hop and not with the ultimate RADIUS server. In other words, the trust relationship between the RADIUS client and the final RADIUS server is transitive rather than using a direct trust relationship. If a server in the chain is compromised, some security problems arise.

- RADIUS does not provide high transport protection. For example, an observer can examine the content of RADIUS messages and trace the content of a specific attribute.

To overcome these security weakness, it has been proposed the use of TLS (T. Dierks & C. Allen (1999)) to provide a means to secure the RADIUS communication between client and server on the transport layer (S. Winter et al. (2010)). Nevertheless, the main research and standardization efforts have focused on the design of a new AAA protocol called *Diameter*.

2.1.2.2 Diameter

Diameter, proposed as an enhancement to RADIUS, is considered the next generation AAA protocol. Diameter is characterized by its extensibility and adaptability since it is designed to perform any kind of operation and supply new needs that may appear in future control access technologies. Another cornerstone of Diameter is the consideration of multi-domain scenarios where AAA infrastructures administered by different domains are interconnected to provide an unified authentication, authorization and accounting framework. For this reason, Diameter is widely used in 3G networks and its adoption is recommended in future AAA infrastructures supporting access control in NGN.

The Diameter protocol defines an extensible architecture that allows to incorporate new features through the design of the so-called *Diameter applications*, which rely on the basic functionality provided by the *base protocol*. The Diameter *base protocol* (P. Calhoun & J. Loughney (2003)), defines the Diameter minimum elements such as the basic set of messages, attribute structure and some essential attribute types. Additionally, the basic specification defines the inter-realm operations by defining the role of different types of Diameter entities. Diameter applications are services, protocols and procedures that use the facilities provided by the Diameter base protocol itself. Every Diameter application defines its own *commands* and *messages* which, in turn, can define new attributes called *Attribute Value Pair* (AVP) or re-use existing ones already defined by some other applications.

The Diameter base protocol does not define any use of the protocol and expects the definition of specific applications using the Diameter functionality. For example, the use of Diameter for providing authentication during network access is defined in the *Diameter NAS Application* (P. Calhoun et al. (2005)). In turn, this specification is used by the *Diameter EAP Application* (P. Eronen et al. (2005)) to specify the procedure to perform the network access authentication by using the EAP protocol. Similarly, authorization and accounting procedures are expected to be handled by specific applications.

Within a Diameter-based infrastructure, the protocol distinguishes different types of nodes where each one plays a specific role:

1. *Diameter Client*: represents an entity implementing network access control like, for example, a NAS. The Diameter client issues messages soliciting authentication, authorization or accounting services for a specific user.

2. *Diameter Server*: is the entity that processes authentication, authorization and accounting request for a particular domain. The Diameter server must support the Diameter base protocol and the applications used in the domain.

3. *Diameter Agent*: is an entity that processes a request and forwards it to a Diameter server or to another agent. Depending on the service provided, we can distinguish:

   (a) *Relay agents*: which forward messages based on routing-related attributes and routing tables.

   (b) *Proxy agents*: which act as a relay agent that, additionally, may modify the routed message based on some policy.

   (c) *Redirect agents*: instead of routing messages, they inform the sender about the proper way to route the message.

   (d) *Translation agents*: which perform protocol translations between Diameter and other AAA protocols such as RADIUS.

The different types of nodes exchange Diameter messages that carry information. Instead of defining a message type, Diameter uses the concept of *command* to specify the type of function a Diameter message intends to perform. Because the message exchange style of Diameter is synchronous, each command consists of a request and its corresponding answer. Table 1 provides a brief summary of the main Diameter commands defined in the base protocol specification.

Command	Abbreviation	Description
Capabilities-Exchange- Request /Answer	CER/CEA	Discovery of a peer's identity and its capabilities.
Disconnect-Peer-Request /Answer	DPR/DPA	Used to inform the intention of shutting down the connection.
Re-Auth-Request /Answer	RAR/RAA	Sent to an access device (NAS) to solicit user re-authentication.
Session-Termination-Request /Answer	STR/STA	To notify that the provision of a service to a user has finalized.
Accounting-Request /Answer	ACR/ACA	To exchange accounting information between Diameter client and server.

Table 1. Common Diameter commands

2.2 The Extensible Authentication Protocol (EAP)

The *Extensible Authentication Protocol* (EAP) (B. Aboba et al. (2004)) is a protocol designed by the *Internet Engineering Task Force* (IETF) that permits the use of different types of authentication mechanisms through the so-called *EAP methods* (e.g., based on symmetric keys, digital certificates, etc.). These are performed between an *EAP peer* and an *EAP server*, through an *EAP authenticator* which merely forwards EAP packets back and forth between the EAP peer and the EAP server. From a security standpoint, the EAP authenticator does not take part in the mutual authentication process but acts as a mere EAP packet forwarder.

One of the advantages of the EAP architecture is its flexibility since does not impose a specific authentication mechanism. Additionally, EAP is independent of the underlying wireless access technology, being able to operate in NGNs. Finally, EAP allows an easy integration with existing Authentication, Authorization and Accounting (AAA) infrastructures (B. Aboba et al. (2008) by defining a configuration mode that permits the use of a backend authentication server, which may implement some authentication methods. These advantages have motivated the success of the EAP authentication protocol for network access control in future NGNs.

2.2.1 Components

The EAP protocol consists of request and response messages. Request messages are sent from the authenticator to the peer. Conversely, response messages are sent from the peer to the authenticator. The different messages exchanged during an EAP execution are processed by several components that are conceptually organized in four layers:

- *EAP Lower-Layer.* This layer is responsible for transmitting and receiving EAP packets between the peer and authenticator.

- *EAP Layer.* The EAP layer is responsible for receiving and transmitting EAP packets through the transport layer. The EAP layer not only forwards packets between the EAP transport and peer/authenticator layers, but also implements duplicate detection and packet retransmission.

- *EAP Peer / Authenticator Layer.* EAP assumes that an EAP implementation will support both the EAP peer and the authenticator functionalities. For this reason, based on the code of the EAP packet, the EAP layer demultiplexes incoming EAP packets to the EAP peer and authenticator layers.

- *EAP Method Layer.* An EAP method implements a specific authentication algorithm that requires the transmission of EAP messages between peer and authenticator.

2.2.2 Distribution of the EAP entities

As previously mentioned, an EAP authentication involves three entities: the EAP peer, authenticator and server. Whereas the EAP peer is co-located with the mobile, the EAP authenticator is commonly placed on the *Network Access Server* (NAS) (e.g., an access point or an access router). Depending on the location of the EAP server, two authenticator models have been defined. Figures 2(a) and 2(b) show the *standalone authenticator model* and the *pass-through authenticator model*, respectively. On the one hand, in the standalone authenticator model (Fig. 2(a)), the EAP server is implemented on the EAP authenticator. On the other hand, in the pass-through authenticator model (Fig. 2(b)), the EAP server and the EAP authenticator are implemented in separate nodes.

In order to deliver EAP messages, an *EAP lower-layer* (e.g., IEEE 802.11) is used to transport the EAP packets between the EAP peer and the EAP authenticator. The protocol used to transport messages between the EAP authenticator and the EAP server depends on the authenticator model employed. More precisely, in the standalone authenticator model, the communication between the EAP server and standalone authenticator occurs locally in the same node. In the pass-trough authenticator model, the EAP protocol requires help of an auxiliary AAA protocol such as RADIUS or Diameter.

2.2.3 EAP authentication phases

As depicted in Fig. 3, a typical EAP conversation [1] occurs in three different phases. Initially, in the discovery phase (*Phase 0*), the peer discovers the EAP authenticator near to the peer's location with which it desires to start an authentication process. This phase, which is supported by the specific EAP lower-layer protocol, can be performed either manually or automatically.

---

[1] Without loss of generality, it is assumed an EAP pass-through authenticator model.

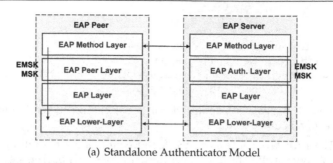

(a) Standalone Authenticator Model

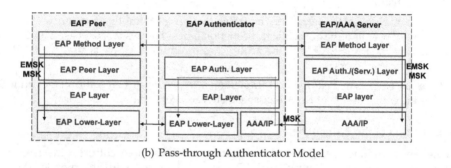

(b) Pass-through Authenticator Model

Fig. 2. EAP authenticator models

The authentication phase (*phase 1*) starts when the peer decides to initiate an authentication process with a specific authenticator. This phase consists of two steps. Firstly, the *phase 1a* includes an EAP authentication exchange between the EAP peer, authenticator and server. To start an EAP authentication, the EAP authenticator usually starts the process by requesting the EAP peer's identity through an *EAP Request/Identity* message. The trigger that signals the EAP authenticator to start the EAP authentication is outside the scope of EAP. Examples of these triggers are the *EAPOL-Start* message defined in IEEE 802.1X (IEEE 802.11 (2007)) or simply an 802.11 association process. On the reception of the *EAP Request/Identity*, the EAP peer answers with an *EAP Response/Identity* with its identity. With this information, the EAP server will select the EAP method to be performed. The EAP method execution involves several exchanges of *EAP Request* and *EAP Response* messages between the EAP server and the EAP peer. A successful EAP authentication finishes with an *EAP Success* message.

Certain EAP methods (Dantu et al. (2007)) are able to generate key material. In particular, according to the *EAP Key Management Framework* (EAP KMF) (B. Aboba et al. (2008)) two keys are exported after a successful EAP authentication: the *Master Session Key* (MSK) and the *Extended Master Session Key* (EMSK). The former is traditionally sent (using the AAA protocol) to the authenticator (*Phase 1b*) to establish a security association with the EAP peer (*Phase 2*). Instead, the latter must not be provided to any other entity outside the EAP server and peer. Thus, both entities may use the EMSK for further key derivation. In particular, as we will analyze in Section 4, some authentication schemes propose to employ the EMSK to derive further key material for enabling a fast re-authentication process.

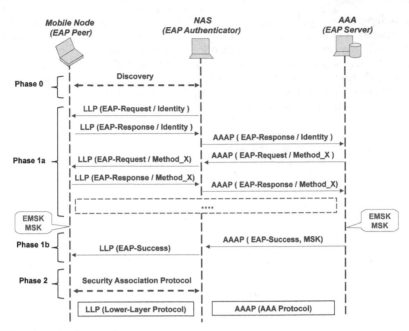

Fig. 3. EAP authentication exchange

2.3 Existing technologies for network access control

The EAP lower-layer protocol allows an EAP peer to perform an EAP authentication process with an authenticator. Basically, the EAP lower-layer is responsible for transmitting and receiving EAP packets between peer and authenticator. Currently, a wide variety of lower-layer protocols can be found since each link-layer technology defines its own transport to carry EAP messages (e.g., IEEE 802.1X, IEEE 802.11, IEEE 802.16e). However, there are also lower-layer protocols operating at network level which are able to transport EAP messages on top of IP (e.g., PANA). Finally, some other lower-layer protocols provide an hybrid solution to transport EAP packets either at link-layer or network layer (e.g., IEEE 802.21 MIH). In the following, the most representative technologies for network access control are analyzed.

2.3.1 IEEE 802.1X

The IEEE 802.1X specification (IEEE 802.1X (2004)) is an access control model developed by the *Institute of Electrical and Electronics Engineers* (IEEE) that allows to employ different authentication mechanisms by means of EAP in IEEE 802 *Local Area Networks* (LANs). As depicted in Fig. 4, there are three main components in the IEEE 802.1X authentication system: *supplicant*, *authenticator* and *authentication server*. In a *Wireless LAN* (WLAN), the supplicant is usually a mobile user, the access point usually represents an authenticator and an AAA server is the authentication server. 802.1X defines a mechanism for port-based network access control. A port is a point through which a supplicant can access to a service offered by a device. The port in 802.1X represents the association between the supplicant and the authenticator. Both the supplicant and the authenticator have a PAE (*Port Access Entity*) that operates the algorithms and protocols associated with the authentication process.

Initially, as depicted in Fig. 4, the authenticator's controlled port is in unauthorized state, that is, the port is *open*. Only received authentication messages will be directed to the authenticator PAE, which will forward them to the authentication server. This initial configuration allows to unauthenticated supplicants to communicate with the authentication server in order to perform an authentication process based on EAP. Once the user is successfully authenticated, the PAE will close the controlled port, allowing the supplicant to access the network service offered by the authenticator's system.

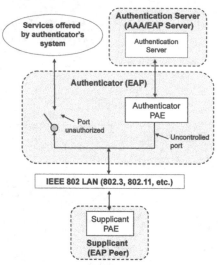

Fig. 4. IEEE 802.1X architecture

2.3.2 IEEE 802.11

IEEE 802.11 extends the IEEE 802.1X access control model by defining algorithms and protocols to protect the data traffic between *station* (STA) and *access point* (AP). More precisely, once the EAP authentication is successfully completed, both STA and AP will share a *Pairwise Master Key* (PMK). This key, derived from the MSK exported by the EAP authentication, is used by a security association protocol (called *4-way handshake*) intended to negotiate cryptographic keys to protect the wireless link between STA and AP. Once the security association is successfully established, the controlled port is closed and access to the network is granted to the supplicant.

The authentication process, described in Fig. 5, involves three entities: an STA acting as supplicant, an AP acting as authenticator and an authentication server (e.g., an AAA server) that assists the authentication process. The process starts with the so-called *IEEE 802.11 association phase* where the STA firstly discovers the security capabilities implemented by the AP *(1)*. Next, the IEEE 802.11 authentication exchange *(2)* is invoked in order to maintain backward compatibility with the IEEE 802.11 state machine. This exchange is followed by an association process *(3)* where the negotiation of the cryptographic suite used to protect the traffic is performed.

In the subsequent *IEEE 802.11 authentication phase*, an EAP authentication is performed where the STA acts as *EAP peer* and the AP acts as *EAP authenticator (4)*. Conversely, the *EAP*

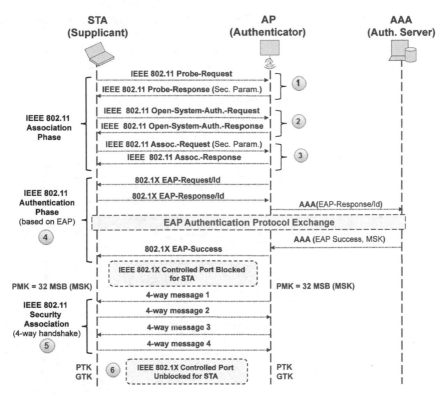

Fig. 5. IEEE 802.11 message flow

server can be co-located with the EAP authenticator (*standalone configuration*) or within an external authentication server (*pass-through configuration*), in which case an AAA protocol (e.g., RADIUS or Diameter) is used to transport EAP messages between the authenticator and the server. Once the EAP authentication is successfully completed, the 32 *more significant bytes* (MSB) from the exported MSK is used as PMK.

Following the establishment of the PMK, a *4-way handshake* protocol is executed during the *IEEE 802.11 security association phase (5)* to confirm the existence of the PMK and selected cryptographic suites. The protocol generates a *Pairwise Transient Key* (PTK) for unicast traffic and a *Group Transient Key* (GTK) for multicast traffic. Thus, as result of a successful *4-way handshake*, a secure communication channel between the STA and the AP is established for protecting data traffic in the wireless link.

2.3.3 IEEE 802.16e

The IEEE 802.16e (*IEEE 802.16e* (2006)) specification is an extension for IEEE 802.16 networks that enables the mobility support and enhances the basic access control mechanism defined for fixed scenarios in order to provide authentication and confidentiality in IEEE 802.16-based wireless networks. In particular, the security architecture is further strengthened by introducing the *Privacy and Key Management* protocol version 2 (PKMv2) which provides mutual authentication and secure distribution of key material between the IEEE 802.16

subscriber station (SS) and the *base station* (BS). The authentication can be performed by using an EAP-based authentication scheme.

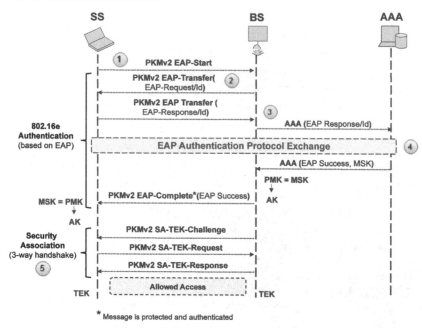

Fig. 6. IEEE 802.16e message flow

Figure 6 shows the authentication process. As observed, while the SS acts as *EAP peer*, the BS implements the *EAP authenticator* functionality. Depending on the EAP configuration mode, the *EAP server* can be placed in the BS (*standalone mode*) or in a AAA server (*pass-through*), which is the case assumed in Fig. 6. As observed, while EAP messages exchanged between SS and BS are transported within the *PKMv2 EAP-Transfer* message, an AAA protocol (e.g., RADIUS or Diameter) is used to convey EAP messages between the BS and the AAA server.

Once the EAP authentication is successfully completed, from the exported MSK a *Pairwise Master Key* (PMK) is derived. In turn, from this PMK, an *Authorization Key* (AK) is generated for the security association establishment. For this reason, the 802.16e specification requires the use of EAP methods exporting key material. Finally, as previously mentioned, the AK shared between SS and BS is employed by a security association protocol called *3-way handshake (5)*, which verifies the possesion of the AK and generates a *Traffic Encryption Key* (TEK) used to protect the traffic in the wireless link.

2.3.4 PANA

The *Protocol for carrying Authentication for Network Access* (PANA) (D. Forsberg et al. (2008)) is a network-layer transport for authentication information designed by the IETF *PANA Working Group* (PANA WG). PANA is designed to carry EAP over UDP to support a variety of authentication mechanisms for network access (thanks to EAP) as well as a variety of underlying network access technologies (thanks to the use of UDP). As highlighted in Fig. 7, PANA considers a network access control model integrated by the following entities:

- The *PANA Client* (PaC) is the client implementation of PANA. This entity resides on the subscriber's node which is requesting network access. The PaC acts as EAP peer according to the EAP model described earlier.

- The *PANA Authentication Agent* (PAA) is the server implementation of PANA. A PAA is in charge of communicating with the PaCs for authenticating and authorizing them to access the network service. The PAA acts as EAP authenticator.

- The *Enforcement Point* (EP) refers to the entity in the access network in charge of inspecting data traffic of authenticated and authorized subscribers. Basically, the EP represents a point of attachment (e.g., access point) to the network.

- The *Authentication Server* (AS) is in charge of verifying the credentials provided by a PaC through a PAA. The AS functionality is typically implemented by an AAA server, which also integrates the EAP server.

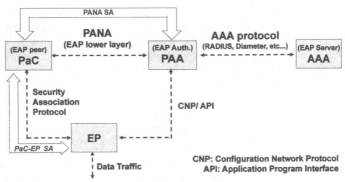

Fig. 7. PANA architecture

Additionally, there are two types of security associations related to PaC in the PANA architecture. On the one hand, a *PANA security association* (PANA SA) is established between the PaC and PAA in order to integrity protect PANA messages. On the other hand, a *PaC-EP SA* is established by performing a security association protocol between the PaC and an EP to protect data traffic.

The PANA operation is developed along four different phases. Initially, during the *authentication and authorization phase*, the PaC and the PAA negotiate some parameters, such as the integrity algorithms used to protect PANA messages. They also exchange PANA messages transporting EAP to perform the authentication and establish a so-called *PANA session*. If the PaC is successfully authenticated, the protocol enters in the *access phase* where the PaC can use the network service by just sending data traffic through the EP. If the PANA session is about to expire, typically a *re-authentication phase* happens to renew this session lifetime. Finally, the PaC or PAA can terminate the session (e.g., the PaC desires to log out the network access session) during *termination phase*, where resources allocated by the network for the PaC are also removed. If neither PaC nor PAA can complete the termination phase, both entities can release the resources once the PANA session lifetime expires.

During each phase, a different set of messages can be sent. Basically we can find four types of PANA messages.

- *PANA-Client-Initiation* (PCI). This message is sent by the PaC requesting the PAA start the authentication process.

- *PANA-Auth-Request/Response* (PAR/PAN). These messages are used during the authentication and authorization phase and the re-authentication phase. They allow to negotiate some parameters between the PaC and the PAA and to carry authentication information in the format of EAP packets.
- *PANA-Notification-Request/Response* (PNR/PNA). These messages are exchanged once PaC is authenticated. They are used as keep-alive mechanism of the PANA authentication session or to signal the beginning of a re-authentication process.
- *PANA-Termination-Request/Response* (PTR/PTA). These messages are used to end up a PANA session.

2.3.5 IEEE 802.21 MIH

The IEEE 802.21 is a recent effort that aims at enabling seamless service continuity among heterogeneous networks (IEEE 802.21 (2008); Taniuchi et al. (2009)). The standard defines a logical entity, *MIH Function* (MIHF), which facilitates the mobility management and handover process. The MIHF is located within the mobility management protocol stack of a mobile node (MN) or network entity. Through the media independent interface, MIHF supports useful services (events, commands or information) that help in determining the need for initiate a handoff or selecting a candidate network

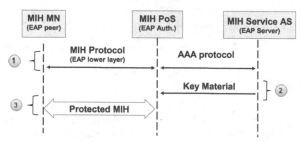

Fig. 8. MIH protocol as EAP lower-layer

Different *tasks groups* (TG) have defined extensions to IEEE 802.21. For example, the standardization task group IEEE 802.21a is defining mechanisms that allow to protect the IEEE 802.21 MIH protocol messages. The solution (EAP over MIH (2010)) designed by the task group proposes that the *mobile node* (MN) must be authenticated and authorized before granting access to the services offered by the *Point of Service* (PoS). In particular, EAP has been proposed as one alternative to carry out this authentication process. Figure 8 depicts the general process followed to perform an EAP-based *Media-Independent Authentication Process*. As observed, the MN and PoS acts as EAP peer and authenticator, respectively. The EAP server functionality is implemented by an entity named *Service Authentication Server* (Service AS). Initially, an EAP authentication (1) is performed between the MN and the Service AS through the PoS, which acts as authenticator. While the MIH protocol is used as EAP lower-layer to transport EAP messages between MN and PoS, an AAA protocol is employed between PoS and Service AS for the same purpose. Note that, since MIH protocol is independent from the underlying transport, this is an hybrid solution that can operate either at link-layer or network-layer. When the EAP authentication is completed, the Service AS sends the MSK (2) exported by the EAP method to the PoS. From this MSK, a key hierarchy is generated to protect MIH protocol packets (3).

3. Fast re-authentication to optimize the network access control

As we can observe, EAP is a promising authentication protocol to be used in NGNs due to its flexibility, wireless technology independence and integration with AAA infrastructures. Furthermore, it is used by a wide variety of network access technologies as standard solution for authentication. However, EAP has shown some drawbacks when mobility is taken into consideration. The reason why the EAP authentication process is not so optimized for mobile scenarios is due to two main motives. First, a typical EAP authentication requires several message exchanges between EAP peer and server. Depending on the EAP method in use (R. Dantu et al. (2007)), this number can vary. For example, one of the most common methods, EAP-TLS (D. Simon et al. (2008)), involves in the best case up to eight messages between peer and server to complete. Secondly, each round-trip is performed with the EAP server placed on the EAP peer's home domain, where the peer is subscribed to. Especially in roaming scenarios, the EAP server may be far from the mobile user (EAP peer) and, therefore, the latency introduced per each exchange increases. These issues are raised when an EAP peer moves from one authenticator to another (*inter-authenticator handoff*). In this case, the peer needs to perform an EAP authentication with the EAP server, through the new EAP authenticator. Therefore, every time the EAP peer moves to a new EAP authenticator, it may suffer from high handoff latency during EAP authentication.

This problem can affect the on-going communications since the latency introduced by the EAP authentication during the handoff process may provoke a substantial packet loss, resulting in a degradation in the service quality perceived by the user. In this sense, the performance requirements of a real-time application will vary according to the type of application and its characteristics such as delay and packet-loss tolerance. The ITU-T G.114 recommendation (ITU-T Recommendation G.114 (1998)) indicates, for Voice over IP applications, an end-to-end delay of 150 ms as the upper limit and rates 400 ms as a generally unacceptable delay. Similarly, a streaming application has tolerable packet-error rates ranging from 0.1 to 0.00001 with a transfer delay of less than 300 ms. As has been proved in (R. M. Lopez et al. (2007)), a full EAP authentication[2] based on a typical EAP method such as EAP-TLS can provoke an unacceptable handoff interruption of about 600 milliseconds (or even in some cases several seconds) for these kind of applications.

To solve this problem, it is necessary to define a *fast re-authentication process* (T. Clancy et al. (2008)) to reduce the authentication time required by a user to complete an EAP-based authentication. Researchers have not ignored this challenging aspect and a wide set of fast re-authentication mechanisms can be found in the literature. Before analyzing the different fast re-authentication schemes in next Section 4, we are going to present both the desired design and security goals that a proper fast re-authentication mechanism should accomplish. To be aware of these requirements is useful to determine advantages and disadvantages when analyzing the different fast re-authentication solutions.

3.1 Design goals

A suitable fast re-authentication solution should accomplish the following requirements and aims (T. Clancy et al. (2008)):

---

[2] Note that the term *full* is used in comparison with *reduced* to denote that, in the execution of an EAP method, there is no optimization to reduce the number of exchanges during the EAP authentication.

(D1) *Low latency operation.* The fast re-authentication mechanism must reduce the authentication time executed during the network access control process compared with a traditional full EAP authentication. Furthermore, the achievement of a reduced handoff latency must not affect the security of the authentication process.

(D2) *EAP lower-layer independence.* Any keying hierarchy and protocol defined must be independent of the lower-layer protocol used to transport EAP packets between the peer and the authenticator. In other words, the fast re-authentication solution must be able to operate over heterogeneous technologies, which is the expected scenario in NGNs. Nevertheless, in certain circumstances, the fast re-authentication mechanism could require some assistance from the lower layer protocol.

(D3) *Compatibility with existing EAP methods.* The adoption of a fast re-authentication solution must not require modifications to existing EAP methods. In the same manner, additional requirements must not be imposed on future EAP methods. Nevertheless, the fast re-authentication solution can enforce the employment of EAP methods following the *EAP Key Management Framework* (B. Aboba et al. (2008)).

(D4) *AAA protocol compatibility and keying.* Any modification to the EAP protocol itself or the key distribution scheme defined by EAP, must be compatible with currently deployed AAA protocols. Extensions to both RADIUS and Diameter to support these EAP modifications are acceptable. However, the fast re-authentication solution must satisfy the requirements for the key management in AAA environments (B. Aboba et al. (2008); R. Housley & B. Aboba (2007)).

(D5) *Compatibility with other optimizations.* The fast re-authentication solution must be compatible with other optimizations destined to reduce the handoff latency already defined by other standards.

(D6) *Backward compatibility.* The system should be designed in such a manner that a user not supporting fast re-authentication should still function in a network supporting fast re-authentication. Similarly, a peer supporting fast re-authentication should still operate in a network not supporting the fast re-authentication optimization.

(D7) *Low deployment impact.* In order to support the aforementioned design goals, a fast re-authentication solution may require modifications in EAP peers, authenticators and servers. Nevertheless, in order to favour the protocol deployment, the required changes must be minimized (ideally, they should be avoided) in current standardized protocols and technologies.

(D8) *Support of different types of handoffs.* The fast re-authentication mechanism must be able to operate in any kind of handoff regardless of whether it implies a change of technology (intra/inter-technology), network (intra/inter-network), administrative domain (intra/inter-domain) or type of security required by the authenticator (intra/inter-security).

3.2 Security goals

In addition to the aforementioned design goals, a secure fast re-authentication mechanism should accomplish the following security goals (R. Housley & B. Aboba (2007)):

(S1) *Authentication.* This requirement mandates that a management and key distribution mechanism must be designed to allow all parties involved in the protocol execution to authenticate every entity with which it is communicating. That is, it must be feasible to gain assurance that the identity of the another entity is as declared, thereby preventing impersonation. To carry out the authentication process, it is necessary to define the so-called *security associations* between the involved entities.

(S2) *Authorization.* During the network access control process, the user is not only authenticated but also authorized to access the network service. The authorization decision is taken by the AAA server and the result is communicated to the authenticator. The fast re-authentication solution proposed must not hinder the authorization process performed once the user is successfully authenticated.

(S3) *Key context.* This requirement establishes that any key must have a well-defined scope and must be used in a specific context for an intended use (e.g., cipher data, sign, etc.). During the time a key is valid, all the entities that are authorized to have access to the key must share the same key context. In this sense, keys should be uniquely named so that they can be identified and managed effectively. Additionally, it must be taken into account that the existence of a hierarchical key structure imposes some additional restrictions. For example, the lifetime of lower-level keys must not exceed the lifetime of higher-level keys.

(S4) *Key freshness.* A key is fresh (from the viewpoint of one party) if it can be guaranteed to be recent and not an old key being reused for malicious actions by either an attacker or unauthorized party (A. Menezes et al. (1996)). Mechanisms for refreshing keys must be provided within the re-authentication solution.

(S5) *Domino effect.* In network security, the compromise of keys in a specific level must not result in compromise of other keys at the same level or higher levels that were used to derive the lower-level keys. Assuming that each authenticator is distributed a key to carry out the fast re-authentication process, a key management solution respecting this property will be resilient against the *domino effect* (R. Housley & B. Aboba (2007)) attack, so the compromise of one authenticator must not reveal keys in another authenticators.

(S6) *Transport aspects.* The solution developed must be independent of any underlying transport protocol. Depending on the physical architecture and the functionality of the involved entities, there may be a need for multiple protocols to perform the transport of keying material between entities involved in the fast re-authentication architecture. As far as possible, protocols already designed and used should be used to address the cryptographic material distribution. For example, while AAA protocols can be considered for this purpose between the EAP authenticator and server, the EAP protocol can be used between EAP peer and server.

4. Overview of existing fast re-authentication schemes

This section analyzes the different efforts that have attempted to reduce the EAP authentication time during the network access control process. According to the strategy followed to achieve this objective, the different fast re-authentication solutions can be classified in different groups: *context transfer, pre-authentication, key pre-distribution, use of a local server* and *modifications to EAP*. In the following, we delve into each of them and detail the mechanism proposed to achieve a reduced handoff latency.

4.1 Context transfer

As depicted in Fig. 9, the context transfer mechanism (T. Aura & M. Roe (2005), H. Kim et al. (2005), C. Politis et al. (2004), *IEEE 802.11 IAPP* (2003), J. Bournelle et al. (2006)) tries to reduce the time devoted to network access control by transferring cryptographic material (1) from an EAP authenticator (*current*) to a new one (*target*). When the user moves to the new authenticator (2), it can use the transferred context (e.g., cryptographic keys and associated lifetimes) to execute a security association protocol with the new authenticator (3) to protect the wireless link. Thus, the user does not need to be authenticated and can directly start the security association establishment process based on the transferred cryptographic material.

In order to perform a secure transference between both authenticators, it is assumed the existence of a pre-established security association between them. Additionally, context transfer solutions do not propagate the same cryptographic material (*CM*) from one authenticator to another. Instead, the transferred cryptographic material is derived (*CM'*) from that owned by the current authenticator where the user is connected. The process employed to generate the derived cryptographic material is followed by both the peer and the authenticator. While the authenticator transfers the derived material to the new authenticator, the peer employs it to start the security protocol execution.

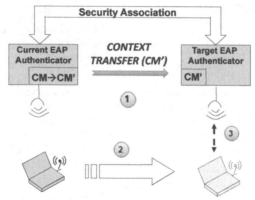

Fig. 9. Context transfer mechanism

Depending on when the transference is performed, we can distinguish between *reactive* and *proactive* schemes. In the proactive mode, the context transfer is performed before the peer performs the handoff. Therefore, when the peer moves to the new authenticator, the cryptographic material has been already transferred to the new authenticator and the peer can immediately establish the security association. Conversely, in the reactive mode, the context transfer is performed once the user performs the handoff and is under the coverage area of the new authenticator. The proactive mode introduces less latency to network access control than the reactive mode since the transference of cryptographic material is performed in advance before the handoff. Nevertheless, reactive solutions are interesting in situations where the handoff happens unexpectedly and there is no anticipation to perform the transference.

An important advantage of context transfer mechanisms relies on their ability to re-authenticate the user without the need of contacting an authentication server located in the infrastructure. Nevertheless, they have been widely criticized as a promising technique to achieve a fast network access due to an important security vulnerability known as the *domino effect* (R. Housley & B. Aboba (2007)). The problem comes from the fact that context transfer re-uses the same cryptographic material (or a derived one following a well-known process) in different authenticators. Therefore, if one authenticator is compromised, the rest of authenticators visited by the same user are also affected.

4.2 Pre-authentication

Pre-authentication solutions propose a scheme (see Fig. 10) where the mobile user performs a full EAP authentication (*1*) with a candidate authenticator through the current associated one *before* it performs the handoff. In this manner, when the handoff happens (*2*), given that the MSK generated during the pre-authentication process will be already present in the candidate authenticator, the peer only needs to establish a security association (*3*) with it to protect the wireless link. As we see, pre-authentication decouples the authentication and network access control operations from the handoff.

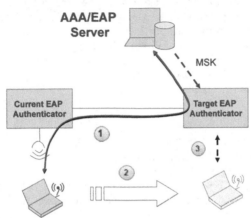

Fig. 10. Pre-authentication mechanism

Depending on the role adopted by the current authenticator during the EAP pre-authentication, we can distinguish two scenarios of EAP pre-authentication signalling (Y. Ohba et al. (2010)):

- *Direct pre-authentication*. In this type of EAP pre-authentication, the current authenticator only forwards the EAP lower-layer messages between mobile node and candidate authenticator as it would be data traffic.

- *Indirect pre-authentication*. Here, the current authenticator plays an active role during pre-authentication process. This type of pre-authentication is useful when the mobile node neither has the candidate authenticator address nor is able to access to the candidate authenticator for security reasons. Therefore, there is a signalling from mobile node to/from current authenticator, and from/to the current authenticator to/from the candidate authenticator. Note that current authenticator does not act as an EAP authenticator; it only translates between different EAP lower-layer protocols.

The first pre-authentication proposal was initially introduced at link layer by the IEEE 802.11i technology (IEEE 802.11i (2005)) and later improved in IEEE 802.11r (IEEE 802.11r (2005)). Nevertheless, the definition of pre-authentication mechanisms at link-layer has some serious limitations since they cannot be applied for cases involving inter-domain or inter-technology handoffs. To avoid this problems, some other solutions propose a pre-authentication procedure at network layer. Network layer solutions (Y. Ohba and A. Yegin (2010), R. M. Lopez et al. (2007), A. Dutta et al. (2008)) have the advantage of being capable to work independent of the underlying access technologies and with authenticators located in different networks or domains.

Despite pre-authentication solutions can potentially achieve an important reduction in the latency introduced by the authentication process during the network access control, this technique presents some drawbacks. First, pre-authentication requires the existence of network connectivity to carry out the pre-authentication process which is a requisite that may not always be satisfied. Second, pre-authentication requires a precise selection of the authenticator with which perform a pre-authentication process. If the user performs a pre-authentication with authenticators where the user finally does not move, the technique may incur in an unnecessary use of network resources. The third disadvantage is related to the previous one. Since pre-authentication implies the pre-reservation of resources in candidate authenticators, in practice, operators are reluctant to pre-reserve resources for users that may or may not roam in the future. Therefore, pre-authentication may have a limited application, specially in inter-domain handoffs. Finally, given that pre-authentication involves a full EAP authentication, special care must be taken to determine the moment to start the pre-authentication process. As a consequence, pre-authentication needs to be performed with a considerable anticipation to the handoff.

4.3 Key pre-distribution

Key pre-distribution solutions (A. Mishra et al. (2004), S. Pack & Y. Choi (2002), Z. Cao et al. (2011), F.Bernal-Hidalgo et al. (2011)) propose the pre-installation of cryptographic material (e.g., keys) in candidate authenticators so that the keys required for secure association are already available when the peer moves to the authenticators. As depicted in Fig. 11, the mobile user initially performs an EAP authentication (1) with the AAA server. Once the EAP authentication is successfully completed, the AAA server pre-distributes keys (2) to authenticators which the user can potentially associate to in a near future. Therefore, when the peer moves to a new authenticator (3 and 5), it is not required to perform a full EAP authentication. Instead, using the key material already present in the authenticator and known by the peer, a security association is established between both entities (4 and 6).

Fast re-authentication solutions based on key pre-distribution have two main disadvantages. On the one hand, they require a precise selection of those authenticators to which pre-distribute key material. If the user pre-distributes key material to authenticators where the user finally does not move, the technique may incur in an unnecessary use of resources. Nevertheless, this is a complex problem given the difficulty of predicting future movements of the user. On the other hand, key pre-installation solutions have a significant deployment cost since a modification in existing lower-layer technologies and AAA protocols is required in order to allow pushing a key provided by an external entity instead of being produced as a consequence of a successful EAP authentication executed through the EAP authenticator.

4.4 Use of a local server

According to the EAP authentication model (B. Aboba et al. (2004)), each time a user needs to be authenticated, a full EAP authentication must be performed with the AAA/EAP server located in the user's home domain. This is a serious limitation for roaming scenarios, specially in mobility contexts. The reason is that each time the visited network needs to re-authenticate the client, the home domain must be contacted. This introduces a considerable latency during network access process since the home EAP server could be located far from the current user's location. Furthermore, taking into account that typical EAP methods (e.g., EAP-TLS) require multiple round trips, the home domain needs to be contacted several times in order to complete the EAP conversation, resulting in unacceptable handoff times.

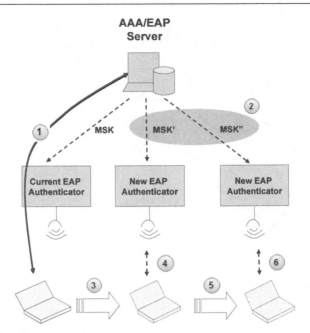

Fig. 11. Key pre-distribution mechanism

To solve this issue, some solutions (3GPP TS 33.102 V7.1.0 (2006), R. Marin et al. (2006), F.Bernal-Hidalgo et al. (2011), V. Narayanan & L. Dondeti (2008)) have proposed the use of a local server near the area of movement of the peer to speed up the re-authentication. The basic idea is to allow the visited domain to play a more active role in network access control by allowing the home AAA server to delegate the re-authentication task to the local AAA server placed in the visited domain. As depicted in Fig. 12, the user firstly performs a full EAP authentication (1) with the home AAA/EAP server using the *long-term* credentials that the home domain provides to their subscribers. This initial EAP authentication, commonly named *bootstrapping phase*, is performed the first time the user connects to the network. Next, once the EAP authentication is successfully completed, the home AAA/EAP server sends (2) some key material (KM) to the visited AAA/EAP server. This key material, which is used as *mid-term* credential between the mobile and the visited AAA/EAP server, allows to locally perform re-authentication (3, 4) when the peer moves to other authenticators located in the visited domain, thus avoiding AAA signalling with the home AAA/EAP server.

Despite this kind of fast re-authentication solutions do not require to contact the home domain to re-authenticate the user, they do not define any optimization for the re-authentication process with the local server. For example, authors in (R. Marin et al. (2006)) propose the use of an EAP method based on shared secret key like EAP-GPSK which requires two message exchanges with the local authentication server. Another serious disadvantage is found in the process followed to distribute the key that establishes a trust relationship between the peer and the local server. Solutions like (F.Bernal-Hidalgo et al. (2011); R. Marin et al. (2006)) use a two-party model to carry out a key distribution process which involves three entities: peer, local re-authentication server and home AAA/EAP server. Since the use of a two-party model is known to be inappropriate (D. Harskin et al. (2007)) from a security standpoint, a three-party approach is recommended.

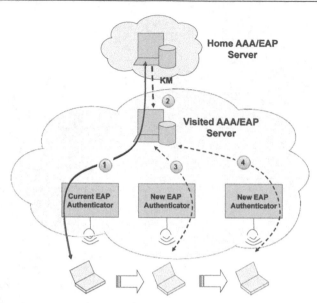

Fig. 12. Use of a local server mechanism

4.5 Modifications to EAP

Finally, another group of solutions try to reduce the EAP authentication time by modifying the EAP protocol itself. Between the different solutions following this approach, the most relevant contribution is the *EAP Extensions for EAP Re-authentication Protocol* (ERP) (V. Narayanan & L. Dondeti (2008)), which has been proposed by the IETF *HandOver KEYing Working Group* (HOKEY WG).

ERP is a method-independent solution that modifies the EAP protocol to achieve a lightweight authentication process. Additionally, ERP relies on the local server optimization (see Section 4.4) and assumes the existence of a local *EAP Re-authentication* (ER) server to optimize the process, which will be in charge of both fast EAP re-authentication and key distribution tasks. The ERP protocol describes a set of extensions to EAP in order to enable efficient re-authentication for a peer that has already established some EAP key material with the EAP server in a previous *bootstrapping phase*. These extensions include three new messages: *EAP-Initiate/Re-auth-Start*, *EAP-Initiate/Re-auth* and *EAP-Finish/Re-auth*.

As shown in Fig. 13, the ERP negotiation involves the peer, the authenticator and the ER server. Beforehand, it is assumed that the peer performs a full EAP authentication with the ER server and both entities share a EMSK. From the EMSK, the peer and the ER server derives a key named rRK. In turn, from the rRK, a new key named *Re-authentication Integrity Key* (rIK) is derived to provide proof of possession and authentication during the re-authentication process.

The ERP re-authentication process is initiated by the authenticator by sending *EAP-Initiate/Re-auth-Start* to the peer. On the reception of this message, the peer sends an *EAP-Initiate/Re-auth* protected with the rIK which is forwarded by the authenticator to the ER server. Once the ER server successfully verifies this messages, it

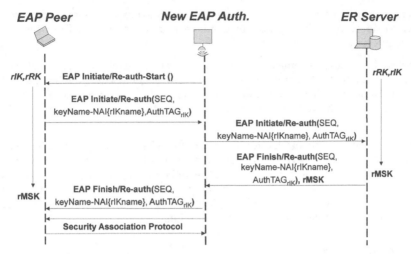

Fig. 13. ERP protocol

replays with a final *EAP-Finish/Re-auth* and derives a rMSK (from the rRK), which is sent to the authenticator to establish a security association with the peer.

On the one hand, in general, the main problem of this kind of proposals relies on their high deployment cost. Since these solutions update the EAP protocol basic operation, they require the modification of existing EAP implementations in order to support the new re-authentication functionality. Consequently, user equipments, authenticators and authentication servers need to be updated, thus complicating the adoption of the solution. On the other hand, in particular, an important drawback of ERP is found on the security of the re-authentication process. Similarly to solutions (F.Bernal-Hidalgo et al. (2011); R. Marin et al. (2006)) previously analyzed in Section 4.4, ERP follows an inappropriate two-party key distribution model to distribute the rMSK from the ER to the authenticator.

5. Conclusion

The provision of *seamless mobility* has created an interesting research field within NGNs in order to find mechanisms which try to provide a continuous access to the network during the handoff. In fact, this is a critical process, where the connection to the network is interrupted, thus causing packet loss that may affect on-going communications. To solve this problem, efforts are directed at reducing the time required to complete the different tasks performed during the handoff. In particular, the network access control process has been demonstrated to be one of the most important factors that negatively affects handoff latency. This process is demanded by network operators in order to control that only legitimate users are able to employ the operator's resources.

This chapter has provided a general overview about the state-of-art of technologies and protocols related to network access control in future NGNs. In particular, we have reviewed the EAP/AAA framework as a promising architecture for network access authentication in future heterogeneous networks. While AAA infrastructures provide an unified framework to handle the authentication, authorization and accounting processes, the EAP protocol is used to implement the authentication service in AAA scenarios. Apart from being easily deployable within existing AAA infrastructures, EAP exhibits important features such as flexibility to select an authentication mechanism and independence from the underlying wireless technology.

Nevertheless, EAP presents some deficiencies when applied in mobile scenarios. In particular, a typical EAP authentication introduces a prohibitive latency during the handoff which provokes a connection disruption that may affect active communications. This problem has been extensively studied by the research community, which has proposed different fast re-authentication mechanisms.

Precisely, the second part of the chapter is devoted to revise and analyze the different schemes that have tried to reduce the latency introduced by network access control during the handoff. According to the strategy followed to reduce the authentication time, we can distinguish five fast re-authentication schemes: context transfer, pre-authentication, key pre-distribution, use of a local server and modifications to EAP. Throughout this chapter we have analyzed both advantages and disadvantages of each approximation.

6. Acknowledgements

This work is partially supported by the Funding Program for Research Groups of Excellence (04552/GERM/06) and the Spanish Ministry of Science and Education (TIN2008-06441-C02-02).

7. References

3GPP TS 33.102 V7.1.0 (2006). 3rd Generation Partnership Project.

A. Dutta, D. Famolari, S. Das, Y. Ohba, V. Fajardo, K. Taniuchi, R. Lopez & H. Schulzrinne (2008). *Media-Independent Pre-Authentication Suppporting Secure Interdomain Handover Optimization*, IEEE Wireless Communications vol. 15(2): 55–64.

A. Menezes, P. van Oorschot & S. Vanstone (1996). Handbook of Applied Cryptography, CRC Press.

A. Mishra, M. Shin, N. Petroni, C. Clancy & W. Arbaugh (2004). *Proactive Key Distribution Using Neighbor Graphs*, IEEE Wireless Communication 11: 26–36.

B. Aboba, D. Simon & P. Eronen (2008). *Extensible Authentication Protocol Key Management Framework*. RFC 5247.

B. Aboba, L. Blunk, J. Vollbrecht, J. Carlson & H. Levkowetz (2004). *Extensible Authentication Protocol (EAP)*. RFC3748.

Badra, M., Urien, P. & Hajjeh, I. (2007). *Flexible and fast security solution for wireless LAN*, Pervasive and Mobile Computing Journal 3: 1–14.

C. de Laat, G. Gross, L. Gommans, J. Vollbrecht & D. Spence (2000). *Generic AAA Architecture*. IETF RFC 2903.

C. Politis, K. Chew, N. Akhtar, M. Georgiades, R. Tafazolli & T. Dagiuklas (2004). *Hybrid multilayer mobility management with AAA context transfer capabilities for all-IP networks*, IEEE Wireless Communications 11 pp. pp. 76–88.

C. Rigney, S. Willens, A. Rubens & W. Simpson (2000). *Remote Authentication Dial In User Service (RADIUS)*. IETF RFC 2865.

D. Forsberg, Y. Ohba, B. Patil, H. Tschofenig & A. Yegin (2008). *Protocol for Carrying Authentication for Network Access (PANA)*. IETF RFC 5191.

D. Harskin, Y. Ohba, M. Nakhjiri & R. Marin (2007). *Problem Statement and Requirements on a 3-Party Key Distribution Protocol for Handover Keying.* IETF Internet Draft, draft-ohba-hokey-3party-keydist-ps-01.

D. Simon, B. Aboba & R. Hurst (2008). *The EAP-TLS Authentication Protocol.* IETF RFC 5216.

Dantu, R., Clothier, G. & Atri, A. (2007). EAP Methods for Wireless Networks, *Computer Standards Interfaces* 29(3): 289–301.

EAP over MIH (2010). Option III: EAP to conduct service authentication and MIH packet protection (21-10-0078-08-0sec-option-iii-eap-over-mih-service-authentication).

F.Bernal-Hidalgo, Marin-Lopez, R. & Gomez-Skarmeta, A. (2011). *Key Distribution Mechanisms For IEEE 802.21-Assisted Wireless Heterogeneous Networks, Mobile Networks and Management*, Vol. 68, Springer Berlin Heidelberg, pp. 123–134.

H. Kim, K. G. Shin & W. Dabbous (2005). *Improving Cross-domain Authentication over Wireless Local Area Networks, Proc. of 1st International Conference on Security and Privacy for Emerging Areas in Communications Networks, SECURECOMM'05*, IEEE Computer Society, Athens, Greece, pp. 103–109.

IEEE 802.11 (2007). Telecommunications and Information Exchange between Systems – Local and Metropolitan Area Network – Specific Requirements – Part 11: Wireless LAN Medium Access Control (MAC) and Physical Layer (PHY) Specifications.

IEEE 802.11i (2005). Std., Wireless LAN Medium Access Control (MAC) and Physical Layer (PHY) Specifications: Specification for Enhanced Security.

IEEE 802.11 IAPP (2003). IEEE Trial-Use Recommended Practice for Multi-Vendor Access Point Interoperability via an Inter-Access Point Protocol Across Distribution Systems Supporting IEEE 802.11 Operation.

IEEE 802.11r (2005). , Wireless LAN Medium Access Control (MAC) and Physical Layer (PHY) Specifications: Amendment 8: Fast BSS Transition.

IEEE 802.16e (2006). Air Interface for Fixed and Mobile Broandband Wireless Access System.

IEEE 802.1X (2004). Standards for Local and Metropolitan Area Networks: Port based Network Access Control, IEEE Standards for Information Technology.

IEEE 802.21 (2008). Institute of Electrical and Electronics Engineers, Draft IEEE Standard for Local and Metropolitan Area Networks: Media Independent Handover Services.

ITU-T Recommendation G.114 (1998). ITU-T General Characteristics of International Telephone Connections and International Telephone Circuits: One-Way Transmission Time, ITU-T Recommendation G.114.

J. Bournelle, M. Laurent-Maknavicius, H. Tschofenig, Y. El Mghazli, G. Giaretta, R. Lopez & Y. Ohba (2006). *Use of Context Transfer Protocol (CXTP) for PANA.* IETF Internet Draft, draft-ietf-pana-cxtp-01.

Marin-Lopez, R., Pereniguez, F., Bernal, F. & Gomez, A. (2010). *Secure three-party key distribution protocol for fast network access in EAP-based wireless networks, Computer Networks* 54: 2651 – 2673.

N. Nasser, A. Hasswa & H. Hassanein (2006). *Handoffs in Fourth Generation Heterogenous Networks, IEEE Communications Magazine* vol. 44(10): pp. 96–103.

P. Calhoun, G. Zorn, D. Spence & D. Mitton (2005). *Diameter Network Access Server Application.* IETF RFC 4005.

P. Calhoun & J. Loughney (2003). *Diameter Base Protocol.* IETF RFC 3588.

P. Eronen, T. Hiller & G. Zorn (2005). *Diameter Extensible Authentication Protocol (EAP) Application.* IETF RFC 4072.

R. Dantu, G. Clothier & Anuj Atri (2007). *EAP methods for wireless networks, Elsevier Computer Standards & Interfaces* vol. 29: pp. 289–301.

R. Housley & B. Aboba (2007). *Guidance for Authentication, Authorization, and Accounting (AAA) Key Management*. IETF RFC 4962.

R. M. Lopez, A. Dutta, Y. Ohba, H. Schulzrinne & A. F. Gomez Skarmeta (2007). *Network-Layer Assisted Mechanism to Optimize Authentication Delay during Handoff in 802.11 Networks*, Proc. of the 5th Annual International Conference on Mobile and Ubiquitous Systems: Computing, Networking and Services, ACM Mobiquitous 2007, ACM, Philadelphia, USA.

R. Marin, J. Bournelle, M. Maknavicius-Laurent, J.M. Combes & A. Gomez-Skarmeta (2006). *Improved EAP keying framework for a secure mobility access service*, Proc. of International Wireless Communications & Mobile Computing Conference 2006, IWCMC 2006, Vancouver, British Columbia, Canada, pp. 183–188.

S. Pack & Y. Choi (2002). *Fast Inter-AP Handoff using Predictive-Authentication Scheme in a Public Wireless LAN*, Proc. of IEEE Networks 2002 (Joint ICN 2002 and ICWLHN 2002).

S. Winter, M. McCauley, S. Venaas & K. Wierenga (2010). *TLS encryption for RADIUS*. IETF Internet-Draft.

T. Aura & M. Roe (2005). *Reducing Reauthentication Delay in Wireless Networks*, Proc. of 1st IEEE Security and Privacy for Emerging Areas in Communication Networks, SECURECOMM 2005, IEEE, Athens, Greece, pp. 139–148.

T. Clancy, M. Nakhjiri, V. Narayanan & L. Dondeti (2008). *Handover Key Management and Re-authentication Problem Statement*. IETF RFC 5169.

T. Dierks & C. Allen (1999). *The TLS Protocol Version 1.0*. IETF RFC 2246.

Taniuchi, K., Ohba, Y., Fajardo, V., Das, S., Yuu-Heng, M. T. C., Dutta, A., Baker, D., Yajnik, M. & Famolari, D. (2009). *IEEE 802.21: Media independent handover: Features, applicability, and realization, IEEE Communications Magazine* 47(1): 112 –120.

V. Narayanan & L. Dondeti (2008). *EAP Extensions for EAP Re-authentication Protocol (ERP)*. IETF RFC 5296.

Y. Ohba and A. Yegin (2010). *Pre-Authentication Support for the Protocol for Carrying Authentication for Network Access (PANA)*. IETF RFC 5873.

Y. Ohba, Q. Wu & G. Zorn (2010). *Extensible Authentication Protocol (EAP) Early Authentication Problem Statement*. IETF RFC 5836.

Z. Cao, H. Deng, Y. Wang, Q. Wu & G. Zorn (2011). *EAP Re-authentication Protocol Extensions for Authenticated Anticipatory Keying (ERP/AAK)*. IETF Internet Draft, raft-ietf-hokey-erp-aak-06.

eTOM-Conformant IMS Assurance Management

M. Bellafkih[1], B. Raouyane[1,2], D. Ranc[3], M. Errais[1,2] and M. Ramdani[2]

[1]*Institut National des Postes et Télécommunications, Rabat,*
[2]*Faculté des Sciences et Techniques, Mohammedia,*
[3]*IT Sud Paris, Evry,*
[1,2]*Morroco*
[3]*France*

1. Introduction

QoS management(Raouyane B. et al., 2009) mechanisms as defined by 3GPP can be viewed as a network-centric approach to QoS, providing a signalling chain able to automatically configure the network to provision determined QoS to services on demand and in real time, for instance on top of a DiffServ-enabled network. However, to envision a deployment of such technology in a carrier-grade context would mean significant further effort. In particular, premium paid-for services with SLA (Service Level Agreement) contracts such as targeted by IMS (Poikselka and Georg, 2009)networks would require additional mechanisms able to provide some degree of monitoring in order to asset the SLAs, while IMS by itself does not provide such mechanisms.

The eTOM (enhanced Telecom Operations Map) (Creaner and Reilly, 2005) functional framework is a widespread reference used to model and analyze networks and services activity. From an eTOM point of view, one could argue that IMS does indeed cover the Fulfilment part of service management, but lacks any means to carry out service Assurance. The eTOM framework proposes a complete set of hierarchically layered processes describing all operator activities in a standard way. It is furthermore sustained by a parallel specification of a standard information model, the SID (Shared Information Data) (TMF GB926 Release 4, 2004). It has to be noted however that both tools, the eTOM and the SID, are generic. Also, the eTOM has been designed at times when Services were viewed as centrally controlled and managed, whereas the IMS is really a distributed layer network.

The work presented in this contribution is an attempt to achieve Assurance functionality for QoS-enhanced IMS services following strictly the eTOM specification, thus filling the functional gap as analyzed earlier; furthermore, two architectures are proposed to be compared: a centralized one and a distributed one.

2. IMS and service provisioning

The composition of the supply chain in NGN network is classically described with three layers. The access layer provides IP (v4 or v6) connectivity regardless of the access technologies (Wireless or Wire-line). The service layer therefore supports technology-agnostic services that are developed independently. The core layer i.e. the control layer is the IMS system which provides the complex signalling responsible for routing sessions between users, invocating   services   and   security-related   tasks   (Figure 1). The information processing and management are carried out by nodes called CSCF (*Call State Control Function*) and HSS (*Home Subscriber Server*). The IMS system introduces a control environment similar to the CS session (*Switched Commutation*) but in CP (*Packet Commutation*).

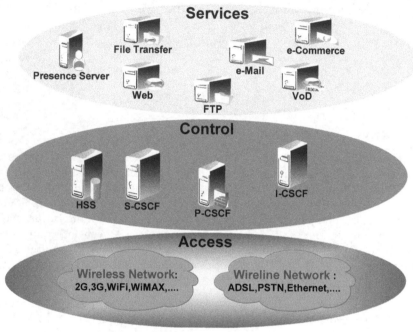

Fig. 1. IMS Layers: Access, Control and Service.

In addition to access unification and diversity of services, IMS introduced a flexible and capable QoS management architecture which organizes exchanges of QoS-related requirements between the control and access layers, allowing resource reservation mechanisms to offer best conditions of supply for e.g. multimedia services.

The service provisioning mechanism of IMS includes three consecutive steps impacting resources: Reservation, Activation and Release (Figure 2).

When a user requests an IMS multimedia service by SIP signalling(Rosenberg et al., 2002) through its attached P-CSCF, the P-CSCF, before forwarding this request must ensure resources availability; this verification is performed through the exchange of Diameter (Korhonen et al., 2010) messages during all media negotiation stages between the two ends (User and AS). An agreement between the client and server can finally lead to change the resource status from reserved into activated. Naturally the PCEF (Policy and Charging Enforcement Function) (3GPP TS 29.210, 2006) applies the relevant QoS policy related to the types of access and transport layers; the most used models are DiffServ(Blake et al., 1998), RSVP(Wroclawski J., 1997) and MPLS (Le Faucheur et al).

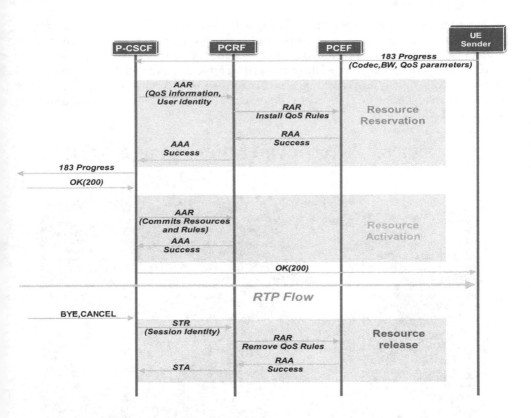

Fig. 2. Service request and negotiation in IMS network with QoS management.

The resources release is carried out at each end of session; the P-CSCF must announce to the PCRF (Policy and Charging Rules Function) (3GPP TR 23.803, 2005) the end of the multimedia session, and the PCRF notifies the PCEF in order to release reserved resources for other applications. QoS management in IMS is a quite flexible on demand mechanism.

3. eTOM (enhanced Telecom Operations Map) architecture

The eTOM is a framework proposed by the TeleManagement Forum and provides a standardized telecom-oriented Business Process map covering all functions of an operator, including service integration and supply. The decomposition layers and functional areas (Customer Service, Resource, and Enterprise) allow detailed operation analysis and to develop solutions according to a well-defined environment. The eTOM has been standardized by the ITU-T (TeleManagement Forum GB921 D, 2010).

The eTOM in its operational part has three main areas: Fulfilment, Assurance and Billing. This section will present only processes related to Assurance, and insist on execution scenarios of SLA (Service Level Agreement)-enhanced services.

3.1 eTOM processes

The 'Operations' area is the traditional heart of the business or service provider (SP). It includes all processes that support client (and network) operations and management. It includes a combination of processes and actions of customer support, including management, provisioning and relationships with partners (Figure 3). The horizontal and vertical processes groupings constitute a matrix formed by a crossing of several processes from level 2, many being derivatives of TOM, which are connected to customer and support operations (FAB).

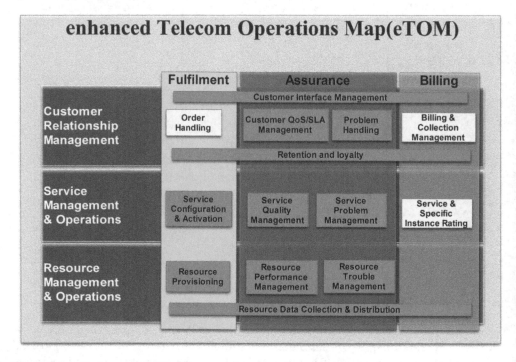

Fig. 3. Operation area in eTOM framework.

A more detailed view of the eTOM business process model (ITU-T Recommendation M.3050.3, 2004) shows a grouping of vertical processes called the FAB columns. These processes are necessary to support operations dedicated to customer satisfaction and operator management:

- **Fulfilment**: Vertical grouping of E2E processes which provide requested services timely and accurately to customers. It reflects business activity. The processes inform customers of their order status, ensure completion on time and customer satisfaction.
- **Assurance**: A group of vertical E2E processes is responsible for implementation of proactive and reactive activities of maintenance to ensure that services are always available and delivered correctly with respect to the SLA. The processes continuously monitor resources status and performance in a proactive way to detect possible defects. They collect performance data and analysis to identify potential problems. In case of

trouble or SLA violation, relevant processes are activated to inform the client about service and trouble status, and to attempt restoration or repair.

- **Billing**: This grouping of vertical E2E process is responsible for collection of appropriate user records, and production of accurate and timely bills, to provide information on resources and services used for payment processing of the customer. In addition, it handles requests from clients on billing, indicates billing status and investigation, and is also responsible for resolving billing issues with respect to customer satisfaction. These processes also support processing of services prepayment.

In addition to the FAB process columns, the Operation area proposes horizontal process groupings:

- **Customer Relationship Management (CRM)**: this group of processes supports knowledge of customer needs and includes all necessary features for acquisition, improvement and maintenance of a relationship with a client. It focuses on service and support, and also on retention management, cross-selling, up-selling and direct marketing. CRM also collect customer and applications information, and customization of service delivery to customers. The processes are responsible for identifying opportunities to increase customer value in company. CRM applies to traditional interactions between client and enterprise.
- **Service Management & Operations (SM&O)**: This group is focusing on services (access, connectivity, content, solution, composition, etc.). It includes all necessary features for management and operations of communications and information required by or proposed to customers. The focus is on service delivery and management of network and information technology. Some functions involve short-term capacity planning service for a service instance, applying a service design to specific customers or managing service improvement initiatives. These functions are closely related to actual experience of customer. The processes in this group are responsible to meet, at a minimum, QoS goals including performance processes and customer satisfaction with service levels and service costs.
- **Resource Management & Operations (RM&O)**: This processes group maintains knowledge of network-related resources (applications, logical and physical infrastructure, communication, management etc.). This group is responsible for managing all these resources (e.g. networks, computer systems, servers, routers, etc.). It is used to provide and support services required by or proposed to customers. The group also contains all features responsible for direct management of these resources (network elements, routers, servers, etc.) used in business process inside operator. These processes are responsible for ensuring that network infrastructure supports an E2E services provisioning. The processes ensure that infrastructure works perfectly, and is available on services and needs and managers.

The R&O group also has a function that allows collection of information from various sources (e.g. network elements (NE) and/or management systems elements (EMS)), and integrates, correlates and in many cases, summarizes data to be transmitted as information relevant to the service management system. This group also includes processes involved in traditional management of network elements (NEM), because these processes are actually essential elements of any process of resource management. RM&O processes thus manage the network service provider and overall infrastructure to ensure reliable interaction with other service providers.

- **Supplier/Partner Relationship Management (S/PRM)**: This process group supports all FAB business processes: Fulfilment, Assurance and Billing. The processes include issuing requisitions and monitoring them until delivery, mediation of requests that must conform to external processes, validating billing and authorizing payment, as well as management quality of suppliers and partners. When an operator sells its products to a partner or supplier, this is done through the CRM business processes, acting on behalf of the supplier or enterprise in such cases.

3.2 System Information & Data (SID)

Naturally the exchange of information between processes is crucial in the eTOM. The detailed specification of the information supporting the eTOM is provided by the SID informational framework (Figure 4). The SID provides an information model capable of interpreting dynamic and static information of business processes and respects the decomposition of the eTOM. The SID specification uses extensively UML class diagrams.

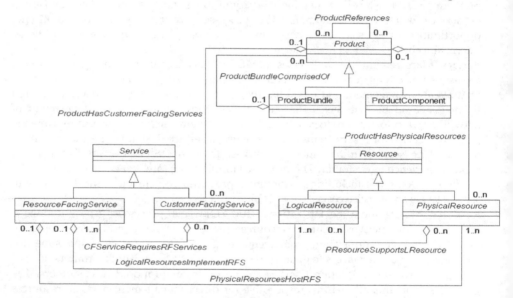

Fig. 4. The main classes of SID: Product, Resource and Service.

- **Product** (TMF GB926 Release 4 Addendum 3 - Product, 2004): in the SID a Product is considered as involving Services and Resources.
- **Service** SID (TMF GB926 Release 4 Addendum 4 SO, 2004): the Product by design is a single or composite service. The Service interacts with the Product to determine its business characteristics, such as customer class and type of service with class *CustomerFacingService*. And the *ResourceFacingService* (TMF GB926 Release 4 Addendum 4 S-QoS, 2004) class exposes resource behaviour for service delivery and its composition for service delivery.
- **Resource**: is divided into two main classes: the **LogicalResource**(TMF GB926 Release 4 Addendum 5 LR, 2004) that exposes logical components and services that are necessary

for service design and product needs; and the **PhysicalResource**(TMF GB926 Release 4 Addendum 5 PR, 2004) which represents physical components of the network such as routers.

3.3 Execution workflows in the eTOM

The eTOM flows during execution scenarios of SLA-monitored service deliveries describe interactions between business processes as well as the information messages that are exchanged in order to handle both cases: the normal execution and the SLA violation.

3.3.1 Normal execution

The normal execution is a normal state of service delivery without SLA violation and the customer will be billed according to services offered and resources reserved. The operation activates a set of processes and many messages are exchanged between them; the SLA verification requires a mapping between Key Performance Indicators (KPIs) and Key Quality Indicators (KQI) related to service and resource instances.

The SLA verification activates a number of separate processes (Figure 5) which are able to assess QoS according to their positions in the different layers: Customer, Service and Resource.

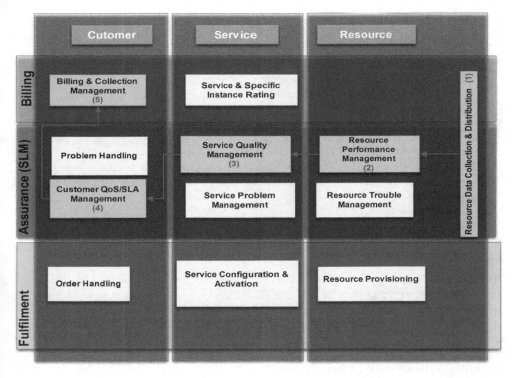

Fig. 5. Active processes in SLA verification.

The SLA verification involves following processes:

- **Resource Data Collection & Distribution**: this process is responsible for the collection of indicators and performance data by contacting all resource agents that provide monitoring, configuration and performance data. The process is also responsible for collecting performance indicators (KPIs) and metrics for all services running in the network. It allows furthermore redistribution of performance data to other processes after aggregation and structuring.
- **Resource Performance Management**: this process reports collected KPIs after filtering and aggregation. The reports provide a structured view of KPIs and a preliminary detection of exceeded thresholds.
- **Service Quality Management**: this process performs a mapping between KPIs and KQIs; it identifies for each service its quality indicators (KQIs) before determining appropriate actions to be performed to calculate them. KQIs values are used to identify failures causes of QoS degradation such a resources failure or lack of capacity in SLA violation.
- **Customer QoS/SLA Management**: is responsible for checking SLA thresholds against measured QoS. After retrieving the KQIs from the Service Quality Management processes and receiving a preliminary report, the process imports the customer profile and SLA parameters to identify thresholds for comparison. It also manages reports of management systems and provides a comprehensive report on the service (metrics, KQIs, key performance indicators, resource use, etc. ...).

The workflow of the SLA verification consists of following steps:

1. When a client requests an IMS service (eg video streaming VoD), the provisioning or "ordering" operation activates all agents in the network to monitor performance indicators and retrieve their values in log files.
2. Resource Data Collection & Distribution retrieves KPIs and metrics collected from different entities in the network. Afterwards, it communicates with the RPM (Resource Performance Management) to identify the existence of critical values and generate performance reports.
3. The performance indicators KPIs collected are sent as XML to Service Quality Management, which identifies indicators KQIs and realize mapping function, and comparing with thresholds are specific to requested service.
4. The Customer QoS / SLA Management uses the loaded profile of customer to identify product thresholds to apply to data collected prior to drafting of audit report of SLA against QoS.
5. The process Billing & Collection Management performs charging functions and taxation with received information to make bills.

3.3.2 SLA violation

The SLA violation scenario begins with a simple verification as above, but in this case a threshold violation occurs. In this case the eTOM provides an escalation mechanism: first, the Resource Layer attempts to solve the problem locally, while warning the Service Layer in order to plan alternative solutions. If the trouble persists, Service processes must perform an alternative service configuration produced by an Ordering operation; this new configuration may be the best solution and is followed by a return to normal SLA verification. The operation chronology consists of three stages: detection and attempted correction, reconfiguration, return to normal verification of SLA.

A real-time continuous monitoring of provided services allows early alerts concerning exceeded thresholds and resource failure alarms, which are main causes of violations and SLA unconformity. Most interactions occur within Assurance processes, but interactions are also concerning the Fulfilment processes, and violation is considered for reimbursement through the Billing processes.

Two specific processes handle the escalation mechanism depicted above: the *Service problem Management* and *Resource Trouble Management* processes (Figure 6).

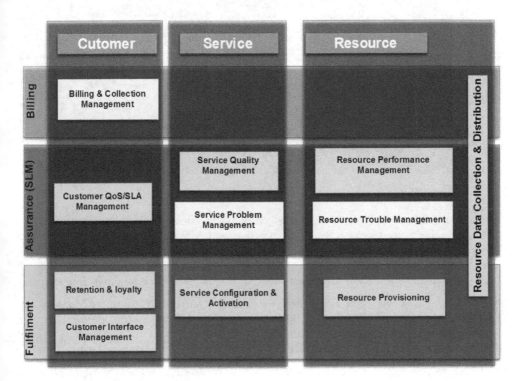

Fig. 6. Active processes in SLA violation.

The goal of these two processes is to perform a restoration of services and resources in short time, and to locate troubles before their expansion, with an optional notification to the user.

The operation is initiated by a usual collection of data by the **RDC&P** process when detecting and exceeded threshold. The process sends relevant information to **RPM** to alert the **RTM** process; in case of a component failure the communication is done directly between **RDC&P** and **RPM**.

The RPM process sends details to the Service Quality Management (SQM) and to the RTM process, depending on the type of trouble, trying to start procedures for resource restoration; for each attempt it notifies the Service Problem Management (SPM) process to synchronize their information about troubles (Figure 7).

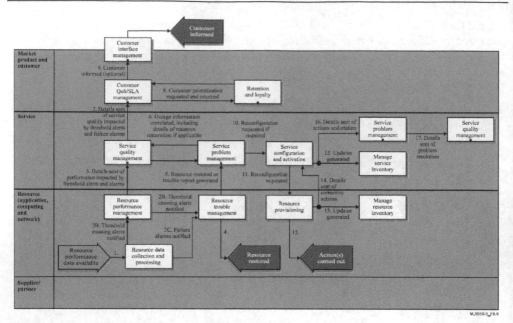

Fig. 7. Processes Flow in SLA execution with violation.

The communication between **SPM** and **SQM** aims to correlate their information about troubles whether solved or not; the **SQM** process sends details of impact on services and alarms to **Customer QoS / SLA Management (CQoS / SLAM)**, a process that specifies SLA violation and customer importance by obtaining all this information from **Retention & Loyalty**, and finally notifies the customer about QoS degradation according to its importance.

If the cooperation between the Service Problem Management and the Resource Management Trouble processes is unsuccessful, the SC&A process will be activated to perform its own corrective action, such as a new configuration. The new configuration will take into account all resource constraints and infrastructure development and service contract terms.

The reconfiguration proposed by SC&A follows exactly the steps of the Ordering operation, and is finalized by launching normal SLA verification, and tries to close all open troubles reports in SPM and RTM. The CQoS / SLAM process can inform the customer about service restoration and quality with the possibility of sending a QoS report.

4. Issues

3GPP specifications provide a basic QoS management architecture for the IMS network which ensures an adequate level of service compared to best effort service. However, the IMS services need to be monitored and managed by a set of mechanisms and methods taking into consideration constraints of the business enterprise. Such a set is explicitly proposed by the eTOM. The eTOM describes its operations and processes in ways that are generic and applicable to any transaction and promises to be fully applicable to the IMS architecture with no applicability constraints. The next step of the study is therefore to plan a mapping strategy in order to map eTOM processes to IMS functions.

5. Functional architecture

A first step in this undertaking is to match IMS functionality with eTOM processes. The resulting set has furthermore to be enriched by eTOM processes relevant to Assurance and Fulfilment. This broader set forms the basis to select different SID entities necessary to carry out these processes. The SLA execution procedure as defined in eTOM model requires the cooperation of several processes belonging to Assurance and Fulfilment of the 'FAB' area, and spanning the three business layers: Customer, Resource, and Service. These eTOM processes will be activated sequentially (Figure 8).

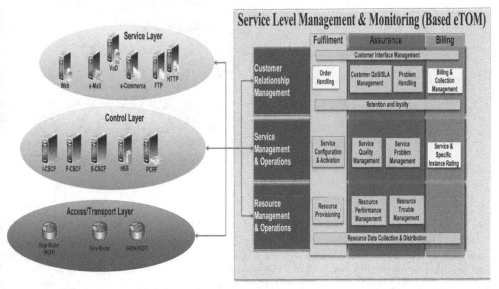

Fig. 8. eTOM and IMS interactions.

The processes belonging to the Assurance layer correspond to the monitoring aspect of this operation, related to Fulfilment for restoration and supply. In order to link eTOM processes to the IMS network, a new component entitled Monitoring, Configuration, Data Collection is required, which clusters the core modules to communicate with these entities.

In the IMS network, the diversity of entities and their various communication protocols require multi-protocol components which can implement all the necessary monitoring and correction operations. An additional constraint is that performance data collection and detection of services should be executed in real time or near real.

5.1 Design

The WSOA (Web Service Oriented Architecture) appears as a valid choice for such a distributed system. The SOA (Mark and Hansen, 2007) concepts will allow to implement EJB (Rima, Gerald, and Micah, 2006) based SOA modules supporting the processes of each component, exposing web services communications via XML/SOAP (Simple Object Access Protocol) /HTTP (Newcomer E., 2002). Three SOA modules have been designed, each of which supporting a part of the targeted eTOM business processes and their associated SID entities. In addition, a BPEL (Business Process Execution Language) (Poornachandra, Matjaz, and Benny, 2006) component has been designed to orchestrate the various processes and to organize the desired operations (Figure 9).

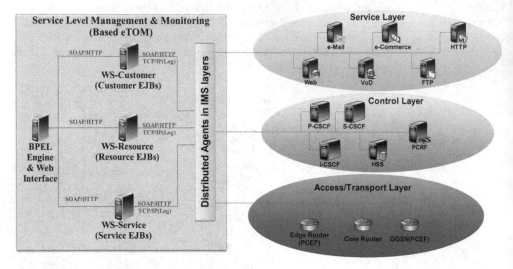

Fig. 9. Implementation architecture.

The three modules of the monitoring system are: WS-Resource, WS-Service, and WS-Customer; each one exposes a set of web services specified using WSDL (Web Services Description Language) (W3C Recommendation, 2007). These web services are invoked and synchronized by the central BPEL component that provides moreover tools such as a web interface that tracks performance of overall network, SLA operation, processes execution and monitoring of physical and logical network resources (Figure 9). The SLM&M (Service Level Monitoring and Management) architecture contain:

- Translation Business Processes: EJB (Enterprise Java Bean) for represent each processes, its functions and information processing.
- Presentation of WSs: WSDL (Web Service Description Language).
- Processes Communication: XML/SOAP (Simple Object Access Protocol).
- Operations Orchestration: BPEL (Business Process Execution Language).
- Communication between SLM&M and the IMS network entities: TCP/IP, XML.

5.2 Centralized architecture

The initial architecture is centralized and enables a selective monitoring of consecutive operations related to SLA and it verification. This system allowed demonstrating the steps of the verification operations, the different KPIs and KQIs of service, and some operational limitations (Raouyane B. et al., 2011).

In order to simplify SLM&M, the number of exposed web services has been limited to eTOM level 3 business processes. Naturally processes of level 4 are implemented via appropriate methods within web services.

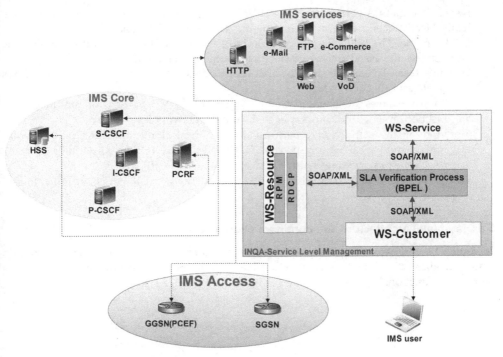

Fig. 10. SLM&M Centralized architecture.

Before analyzing the different SOA modules, it is useful to introduce the interfaces between SOA modules and network. These interface agents take (Figure 10) in charge low-level detections and calculations and transmit their results to the SOA modules via dedicated socket interfaces:

- The IMS agent scans S-CSCF activity and detects service launching ;
- The Application Server agent scans Application Server activity in order to identify the customer parameters ;
- The Router agents perform network analysis tasks in order to calculate KPIs that will be transmitted to SOA modules

The SOA Modules expose each eTOM layer:

- WS-Resource: This SOA module is composed of classes implementing operations defined in the Resource layer of eTOM, as well as corresponding SID entities. It implements two main eTOM processes already discussed in functional architecture: Resource Data Collection & Processing, and Resource Performance Management. Both of them are exposed as web services.
- WS-Service: This module implements various SID entities and operations defined in the Service layer of eTOM. The module exposes processes as web services responsible for quality indicator mapping and analyzing.
- WS-Customer: This module implements functionality defined in the Customer layer of the eTOM and its SID model. It exposes Customer QoS/SLA Management web service

responsible for SLA verification. It retrieves the KQIs of the currently delivered service, loads the customer profile and subsequently detects any SLA violation.

- BPEL Engine: The BPEL Engine module implements a BPEL process that invokes the web services described above and synchronize their interaction.
- Web interface : To monitor the SLA operation and its process, the BPEL engine features a web interface that allows to:
  - Show messages exchanged between web services (XML/SOAP) and modules.
  - List performance indicators collected from the network layer entity
  - Monitor the activity and performance of physical resources such as network routers and logical entities such as CSCFs and the HSS (Home Subscribe Server).
  - Check the provisioning chain of QoS management: PCRF, PCEF.
  - View the results of the audit and SLA verification, the customer class, and values of KQIs.

5.3 Distributed architecture and continuous monitoring

To reduce SLM&M complexity, an enhanced architecture proposes to split the RDC&P into many smaller distributed and decentralized components. Additionally, continuous monitoring functionality has been added to the system (Figure 11).

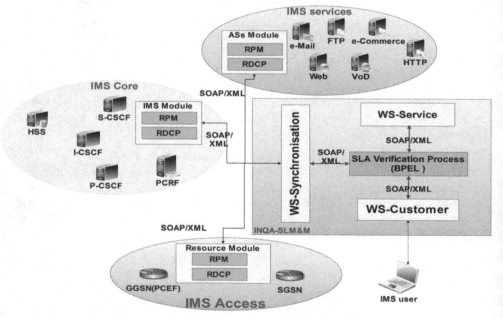

Fig. 11. SLM&M distributed architecture.

However, the centralized web services that allow KPIs and KQIs monitoring are still relying on BPEL technology and still are very resource and time consuming. Indeed, the distribution of the Resource processes allows not only to share processes of KPIs but also to evaluate services locally. Thus, the distribution of EJB modules becomes necessary to incorporate mechanisms for monitoring locally but also to allow a local correction of QoS degradation and anomalies.

The new functional architecture of SLM&M (Figure 12) consists therefore of two main modules:

- **Assurance Layer**: represents the SLA verification process as defined in eTOM in both layer Customer and Service. Thus, processes that are related to operations Fulfilment and Assurance, and the information and data is stored in Customer Inventory and Service Inventory.
- Monitoring Layer:
  - Is distributed, and contains a set of agents and probes that are able to recover all data in real time (signalization, logging, reservation, configuration, policy, routers status, etc. ..) and implements all Resource layer processes for SLA verification: Resource Data Collection & Processing (RDC&P) and Resource Performance Management (RPM), which are related to each IMS layer (Access, Control, Service).
  - Contains a set of processes that are functional in ordering and other SLA operations of WS-Resource. Also, a synchronization module that is necessary for detection and control of events in the network, such as planning activities and communications on one hand between the distributed modules (first part) and also between Web services exposed (Layer).

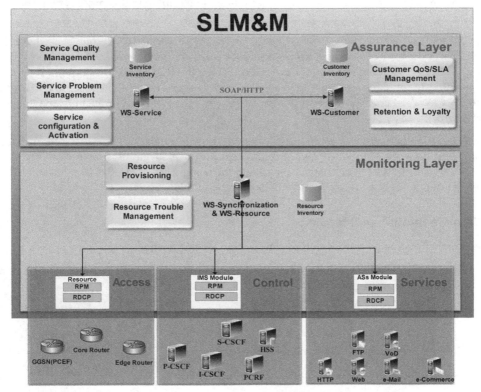

Fig. 12. The main layer of SLM&M: Monitoring, Assurance.

The proposed functional architecture supports three communication channels between different modules:

- TCP / IP between agents and the synchronization module,
- SOAP / XML between layers of eTOM (Resource, Service, Customer) or WSs
- With ability to use XMLCONF between the synchronization module and management in case of SLA violation.

5.4 Correction architecture

The architecture of SLM& M consists of two layers: Assurance and Monitoring, by analogy with the previous architecture (distributed). The Monitoring layer is distributed and contains only two main processes of collection and processing of information.

New processes must be integrated in a centralized way; a distributed integration can overload collection agents in routers and. For example in a router memory is crucial,; collection agents and processes DRC & P and RPM are reasonable for just performance collection and data local treatment. However, the addition of another process could overload the router that needs its capacities for traffic conditioning and processing.

The Resource Trouble Management process (RTM) catches alarms that reflect a degradation of service resulting from a physical or logical related to equipment; this process then tries to make a preliminary correction of the service and notifies WS-Service.

The WS-synchronization process is located in the same server as WS-Resource, so that this server can synchronize incoming events and data collection, and decide either to perform a normal SLA verification, or to report a violation. Also, the Resource Provisioning process is responsible for making resource reservations with respect to solution recommendations provided by WS-Service. WS-Service adds to its repertoire Service Problem Management (SPM) processes.

The interaction between WS components of SLM&M is through SOAP/HTTP, whereas the interaction between Monitoring layer of SLM &M and IMS layers uses Java-based client / server communication, with a spare possibility of using XML/RPC (Mi-Jung et al., 2004) between the (Resource, IMS ASs) and WS-Resource modules.

6. Implementation and results

The implementation encompasses three fundamental components:

- The IMS network for service delivery: control entities (CSCF, HSS) and Application server for video streaming VoD.
- The QoS management: PCEF and PCRF.
- The Monitoring and managing System: SLM&M.

6.1 Trial infrastructure in SLA verification

The trial architecture exposes the function of each component.

The test bed is composed of (Figure 13):

- A core router and two edge routers (Linux boxes) defining a DiffServ-enabled network on which are connected an IMS terminal ad an Application Server;
- This network is controlled by the OpenIMS (Open IMS Core) system which is deployed in the core router Linux box;
- A management server supports the QoS monitoring/Assurance functionality.

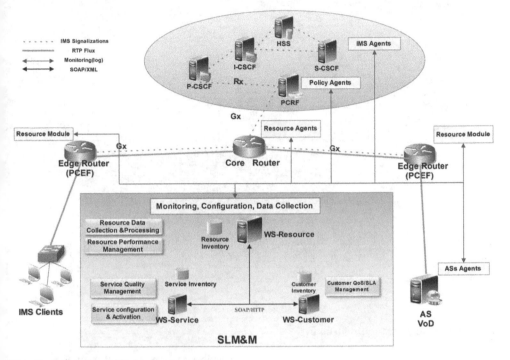

Fig. 13. Centralized trial infrastructure.

6.2 Trial infrastructure for SLA violation

The implementation architecture features two sub-architectures for the service provisioning and for service management and correction services.

The Supply Architecture which contains an IMS network that includes both the signalling and the media planes. The architecture includes three routers to transmit the media stream; a central router supports the IMS system. The PCRF (Policy and Charging Rule Function) is becoming an autonomous entity and includes other features such as policy management, and both edge routers include the PCEF (Policy and Charging Enforcement Function) functionality to receive and execute policies or PCC rules (Figure 14).

Monitoring and management architecture: SLM & M is divided into two layers

- *Monitoring Layer*: contain the two WSs Resource and Synchronization, with the integration of RP and RTM processes and Resource Inventory, so the layer includes the functionality of PCRF for QoS management and control.

- *Assurance Layer*: contain both servers and WS-Customer and WS-Service, and integrate process and Fulfilment and Assurance, that will be activated in SLA correction and violation.

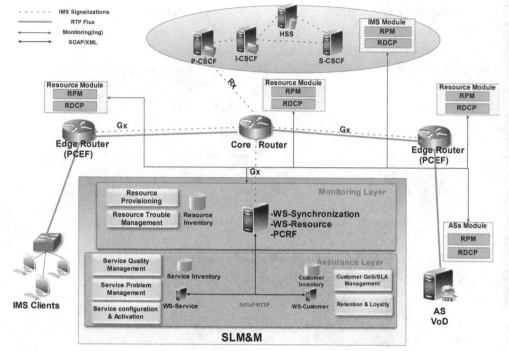

Fig. 14. Centralized trial infrastructure.

The supply architecture provides a set of IMS services, when a client requests a VoD streaming service, the provisioning chain stimulates IMS entities to provide a resource reservation and QoS management. The SLM&M in supply, after ordering operation, start collection of configuration data for a normal SLA execution. Anomaly detection or exceeding threshold causes an activation of SLA violation processes for restoring service into normal level.

The SLM&M must be reactive by rapid detection of QoS degradation or anomalies, followed by an attempt to resolve troubles, that activates the Assurance process and if necessary the Fulfilment process.

6.3 Scenarios

The test scenario includes three cases, a customer Alice with Platinum class that requires a VoD service:

- The client receives service with perfect QoS;
- An overloaded network with a slight QoS degradation;
- Network Congestion and violation of SLA.

6.4 Results

The results expose several parameters relating to monitoring service and performance of SLM&M in centralized and distributed architecture, and response time in trouble detection and SLA violation.

6.4.1 SLA verification in centralized architecture:

Case 1: The QoS offered to Alice and Bob matches SLA contract, perceived video quality is satisfying (Figure 15).

Fig. 15. Video bandwidth =128kbps.

Case 2: the network conditions, hence video quality, deteriorate proportionally to mass of competing services for lost packets and reduced flow rate (Figure 16).

Fig. 16. Video bandwidth =76kbps.

Case 3: competing services overload the routers: queues fill in gateways, impacting delay and jitter. Routers discard packets in excess, this causes static pixels in video (Figure 17) and in some cases service cancellation.

Fig. 17. Video bandwidth =40kbps.

The platform succeeds in identifying accurately deterioration of delivered services. The cost in terms of response time has been evaluated as well. It is observed that response time for Resource-WS is much longer than for other web services, due to complexity of its tasks (Figure9, 10).

Fig. 18. Response time for different web services (Client: Alice).

The number of web services and their internal functionality has a considerable impact on running time of SLA verification. This led to limit the exposed eTOM processes to level 3 and to implement sub processes via internal java methods.

6.4.2 Centralized vs distributed architecture

The execution time in SLA verification is composite, and is directly related to processing time in each WS. This time varies depending on the number of planned operations, WS state and SLM&M conception. Similarly, nature of communication technology between entities plays a vital role in reduction of complexity and processing, which highlights the advantage of using TCP /IP for exchange parameters of service and performance indicators and transmission at higher levels in order to achieve continuous monitoring.

The response time of WS-synchronization and other agents resource is short compared to WS-Resource in SLA verification, because of the processing performed locally in each device, that has a potential to reduce traffic between Monitoring and Assurance layers, as well as it reduces response time compared to centralized architecture in tree cases (Figure 19).

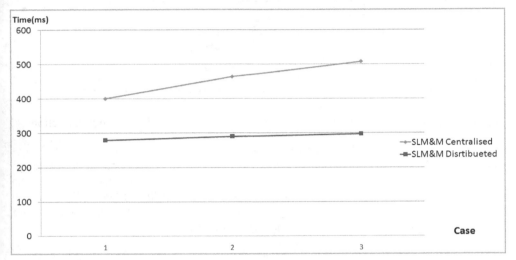

Fig. 19. Response time (ms) for centralized and disturbed architecture.

The processing of parameter flow in entity level allow a real-time control of multiple QoS and services, however a router must have enough memory for traffic conditioning, although using control function as treatment and comparison with thresholds can reduce its capacity in terms of CPU and memory.

The distributed architecture provides a Monitoring layer that alerts Assurance layer within a very short time or near real time, and allows rapid processing of SLA violations, compared to another architecture where several treatments must be executed to detect degradation QoS or failure of an entity.

6.4.3 SLA violation

Alice has registered in the IMS system with QoS classes Gold. The goal is to perform SLA Assurance tests in three representative cases and to compare results for SLA correction with Assurance and Fulfilment.

The MOS-AV (Mean Opinion Score-AudioVisual (BELLAFKIH et al., 2010) is a quality indicator for detecting user satisfaction requiring a video flow such as VoD (Video on demand) or IPTV (IP Television). It is a quantitative indicator taking values between 0 and 5. The MOS-AV reflects user satisfaction and SLA violation in our case the value 2.5 represent a QoS degradation and SLA violation (Figure 20).

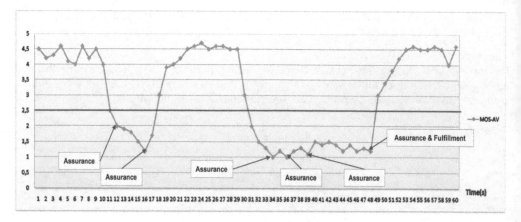

Fig. 20. Customer satisfaction during a time interval after SLA verification and correction.

The MOS-AV indicator reflects customer satisfaction. When detection thresholds are exceeded or values of MOS-V become critical, SLA violation has launched the first attempt with confidence and was successful in restoring normal levels of service after 7 seconds. The second violation that requires intervention of Assurance & Fulfilment takes 17 seconds to restore the service. These results are justified by the architecture that used WSOA and interaction between different WSs and attempted solutions.

7. Conclusion

The proposed approach is based on the QoS provisioning architecture proposed by 3GPP with eTOM Assurance capabilities of QoS monitoring. SLM&M uses the new concepts of SOA and BPEL in managing and monitoring network and also must meet several constraints of instrumentation. The first version of the platform was centralized to address all performance data in a central node, this design offered SLA verification but still remained isolated from IMS network and real events. However the IMS network requires permanent and real time monitoring rather than just a sporadic SLA verification.

The solution to distribute Resource layer and processes of the eTOM and their adjustments to each layer of the IMS (Access, Control, and Service) is appropriate, as the creation of a WS- Synchronization which synchronizes operations process WS-Resource and their agents in network layer, in addition provides operations in the monitoring layer of SLM & M. The distributed architecture demonstrates its ability in terms of response time and is preferable to a centralized SLM&M.

The life cycle of SLM&M has three main stages: a real-time monitoring of services and resources to detect anomalies or degradation, followed by a stage of responsiveness to correct troubles, and the final step is to be proactive in order to estimate the behaviour of service and resource by correlation and root cause analysis of service impact. The proactive property will be integrated with QoS mechanisms that predicted from current data, a mathematical model or stochastic processes that come into the perspective.

8. Acknowledgment

Special thanks are due to MÉDITELECOM operator for its financing, and supporting.

9. References

3rd Generation Partnership Project "Charging rule provisioning over Gx interface (Release 6)", 3GPP TS 29.210 V6.7.0 2006-12, available at http://www.3gpp.org/ftp/Specs/html-info/29210.htm.

3rd Generation Partnership Project, Evolution of policy control and charging (Release 7), 3GPP TR 23.803 V7.0.0 (2005-09), available at http://www.3gpp.org/ftp/Specs/html-info/23203.htm.

Bellafkih, M.; Raouyane, B.; Errais, M.; Ramdani, M.;MOS evaluation for VoD service in an IMS network, I/V Communications and Mobile Network (ISVC), 2010 5th International Symposium, Rabat , Morocco.

Blake S., Black D., Carlson M., Davies E., Wang Z., and Weiss W., An Architecture for Differentiated Services, December 1998, RFC 2475, available at http://www.ietf.org/rfc/rfc2475.txt

Creaner M. J. and Reilly, J. P, NGOSS Distilled: The Essential Guide to Next Generation Telecoms Management. The Lean Corporation 2005.

Enhanced Telecom Operations Map (eTOM) The Business Process Framework for the Information and Communications Services Industry, Addendum D: Process Decompositions and Descriptions Release 6.0 GB921 D; TMF.

J. Wroclawski, The Use of RSVP with IETF Integrated Services, September 1997, RFC 2210, available at http://www.ietf.org/rfc/rfc2210.txt

Korhonen, J., Tschofenig, H., Arumaithurai, M. Jones, M., Ed., and A. Lior, "Traffic Classification and Quality of Service (QoS) Attributes for Diameter", RFC 5777, February 2010, available at http://www.ietf.org/rfc/rfc5777.txt

Le Faucheur F., Wu L., Davie B., Davari S., Vaananen P., Krishnan R., Cheval P., Heinanen J., "Multi-Protocol Label Switching (MPLS) Support of Differentiated Services", RFC 3270, May 2002, available at http://www.ietf.org/rfc/rfc3270.txt.

Mark Hansen, D. (2007) SOA Using Java Web Services, Prentice Hall, New Jersey, USA

Mi-Jung Choi, Hyoun-Mi Choi, Hong, J.W., Hong-Taek Ju, XML-based configuration management for IP network devices, Communications Magazine IEEE, July 2004, Volume: 42 Issue:7, pages: 84 – 91.

Newcomer, E. (2002) Understanding Web Services: XML, WSDL, SOAP, and UDDI, Addison-Wesley Professional.

OpenIMScore – Open source implementation of IMS Call Session Control Functions and Home Subscriber Service (HSS) -http://www.openimscore.org/

Poikselka M. and Georg M. The IMS: IP Multimedia Concepts and Services, John Wiley & Sons Inc. Chichester, England2009.

Poornachandra, S., Matjaz, J., and Benny, M. (2006) Business Process Execution Language for Web Services BPEL and BPEL4WS, 2nd Edition, Paperback.

Raouyane B., Bellafkih M., Ranc D., QoS Management in IMS: DiffServ Model', Paper Presented at the Third International Conference on Next Generation Mobile Applications, Services and Technologies, 15-18 September 2009. Cardiff, Wales, UK.

Raouyane B.; Bellafkih M.; Errais M.; Ranc D.; WS-Composite for Management & Monitoring IMS Network; *International Journal of Next-Generation Computing (IJNGC)* - ISSN 2229-4678, eISSN 0976-5034.

Rima, P. S., Gerald, B., and Micah, S. (2006) *Mastering Enterprise JavaBeans 3.0*, Paperback.

Rosenberg, J., Schulzrinne, H., Camarillo, G., Johnston, A., Peterson, J., Sparks, R., Handley, M. and E. Schooler, "SIP: Session Initiation Protocol", RFC 3261, June 2002, available at http://www.ietf.org/rfc/rfc3621.txt

SERIES M: Telecommunications management network Enhanced Telecom Operations Map (eTOM) –Representative process flows, ITU-T Recommendation M.3050.

Shared Information/Data (SID) Model System View Concepts and Principles, *GB926 Version 1.0, Release 4.0*, January 2004.

Shared Information/Data (SID) Model, Addendum 3 - Product Business Entity Definitions GB922 Version 3.1, NGOSS Release 3.5 July 2003.

Shared Information/Data (SID) Model, Addendum 4SO – Service Overview Business Entity Definitions, GB922 Addendum-4SO, NGOSS Release 3.5 July 2003.

Shared Information/Data (SID) Model, Addendum 4S-QoS Quality of Service Business Entity Definitions, GB922 Addendum – 4S-QoS Version 1.0, NGOSS Release 3.5 July 2003.

Shared Information/Data (SID) Model, Addendum 5LR – LogicalResource Business Entity Definitions, GB922 Addendum-5LR Version 1.0, NGOSS Release 3.5 July 2003.

Shared Information/Data (SID) Model, Addendum 5PR – Physical Resource Business Entity Definitions, GB922 Addendum-5PR Version 3.0 NGOSS Release 3.5 July, 2003.

Web Services Description Language (WSDL) Version 2.0, W3C Recommendation 26 June 2007, http://www.w3.org/TR/wsdl.

3

IP and 3G Bandwidth Management Strategies Applied to Capacity Planning

Paulo H. P. de Carvalho, Márcio A. de Deus and Priscila S. Barreto
Departament of Electrical Engineering, Departament of Computer Science
University of Brasilia
Brazil

1. Introduction

This chapter discusses the application of methodologies to plan and design IP Backbones and 3G access networks for today's Internet world. The recent trend of the multi-frequency band operations for mobile communication systems requires increasingly bandwidth capacity in terms of core and access. The network planning task needs mathematical models to forecast network capacity that match the service demands. As the nature of network usage changed, to explain and forecast the network growth, new methods are needed. In this chapter, we will discuss some strategies to optimize the bandwidth management of a real service provider IP/MPLS backbone and later we will propose a method for traffic engineering in a national IP backbone.

Currently, all telecommunications networks are using IP packets to transport several kind of services. The industry has called this integration as IMS (IP Multimedia Subsystem) in 3G technologies. One important challenge is how to implement this desirable integration with the lack of well known mathematical models to perform capacity planning and forecast the network needs in terms of growth and applications demands. In other way, the main question is how to deliver the required level of service for all kind of applications using the same structure but with different types of traffic and QoS (Quality of Service) requirements.

Due to the fact that many different services will use the same transport infrastructure, the Quality of Service can also be described as a result of traffic characterization because the traffic nature per service or at least per application shall be known. As demonstrated in some research papers (Leland et al., 1994; Carvalho et al., 2009), the Erlang model is not able to accurately describe the behavior of Ethernet and Internet traffic. Without the right model, scientific prediction becomes very difficult and therefore, the planning and forecasting tasks become almost impossible. The above research works verified that the Poisson traffic model is not able to explain the IP traffic dynamics and this implies that the capacity planning tasks for integrated services will need new methodologies. Some models have been used with superior performance to achieve these goals, the self-similar or mono-fractal model show acceptable results in several situations (Carvalho et al., 2007).

Several works show that the multifractal models are particularly promising for multimedia networks (Riedi et al., 2000; Abry, 2002; Fonseca, 2005; Deus, 2007). The traffic engineering task is valuable to optimize the network resources such as links, routing and processing capacity. One important issue in the traffic engineering task is that the capacity planning forecasting may be for medium long periods (or more than one year), due the fact is not easy to increase long distance link capacities in small periods of time. This problem is much more valuable when the coverage area income is not proportional to the area, as in countries like Brazil, China, Russia in which large areas not necessarily economically attractive.

2. Network planning

The planning task is fundamental to optimize resource utilization. The Fig. 1 describes, from an industry point of view, a complete feasible telecommunications planning cycle. The inputs are the service demands, described as all type of products/services needs per region and also per customer. The physical and logical inventory are very important to be accurate in terms of transmission mediums such as fiber or radio, demographic dispersion, network elements complete description, management assets, and other important physical and logical information.

In terms of innovation, the approach is to use new technologies to achieve new degrees of service delivery; this function shall be used as a complement for planning and forecasting purposes. Other very important function is the economic variables to calculate the return of the investments (ROI) and all other related costs (fixed and variable). All information about traffic usage will be collected and sampled depending on the nature of the service and will have a fast track for immediate operations and decision-making, normally every 5 minutes. For long-term planning these samples will be aggregate in hours, days and weeks.

The functions in Figure 1, in terms of long term capacity will be used to achieve the capacity to deliver new services allowing network expansion related to the inputs, generating new routing and topology and other capacity needs, as described in Figure 1. The traffic engineering function is used in real-time, under human supervision, sometimes even when some modification in terms of routing is proposed by an algorithm. Sometimes, this could not be feasible in practice because network stability is more important in operational environments (Carvalho et al., 2009; Evans & Filsfils, 2007).

The peering agreements will be done as a function of the outputs and also observing the commercial issues. In this way, many service providers have a peering committee to approve new peering interconnections, which has not only a technical importance as well as a marketing approach. The capacity outputs will generate purchasing activities; this will be done by an engineering implementation function. The main objective is to have an operational network, providing all kind of facilities and desirable services.

Along with the massive growth of the Internet and other applications, an increasing demand for different kinds of services for packet switching networks is important. Nowadays, these networks are expected to deliver audio and video transmissions with quality as good as that of a circuit switching network. In order to make it possible, the network must offer high quality services when it comes to bandwidth provisioning, delay, jitter and packet loss.

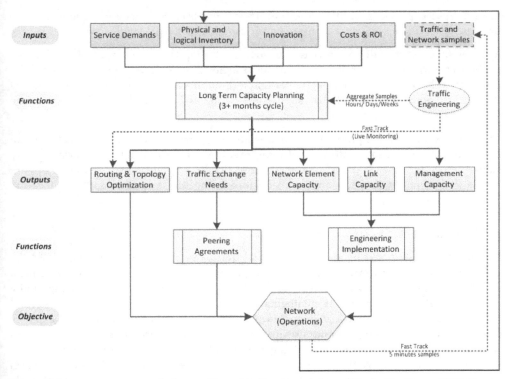

Fig. 1. Telecommunications Industry Planning Process. Adapted from (De Deus, 2007; Evans & Filsfils, 2007).

The processes of traffic characterization and modelling are very important points of a good network project. A precise traffic modelling may allow the understanding of a physical network problem as a mathematical problem whose solution may be simpler. For example, the use of traffic theory suggests that mathematical models can explain, at least for some confidence degrees, the relationship between traffic performance and network capacity (De Deus, 2007; Fonseca, 2005).

The next sections will provide an example on a 3G network using traffic samples to study the planning and project deployment phases. The network described in our study runs with more than 1 million attached 3G costumers with national coverage. In this network, we collected traffic in July 2009 in three different locations (Leblon, Barra da Tijuca and Centro) in Rio de Janeiro. In this way, the first step was to classify the traffic per application. The second step was to characterize the traffic using a procedure based on self-similarity (Clegg, 2005) or multifractal analysis (Carvalho et al., 2009). These results were used as basis for proposing a method to manage the traffic in the network.

To manage the traffic demands, we deployed a traffic engineering concept that divides the traffic across the network through tunnels. The bandwidth was monitored and in the observed period, we collected metrics that were used as inputs to decide how to configure new parameters that may fit the incoming needs. An ILEC (incumbent local exchange carrier) service provider of IP traffic was used to collect real network traces and we simulated a similar architecture of this network using the OPNET Modeler tool.

A 3G with a Metro Ethernet access was also analysed. The analysis considered a per application separation of traffic. The statistical analysis was done using a self-similarity approach, calculating the Hurst parameter using different calculation methodologies (Abry et al., 2002). Some multifractal analysis was also done as a tool to better choose the time scale.

The results show that the proposed method is able to generate better results in terms of an on-line traffic engineering control and also to provide key information to long term capacity planning cycles. The Traffic Engineering function is detailed using some network simulations examples. Finally, some long term forecasting and short term traffic engineering proposal was done in a 3G networks.

2.1 Traffic modelling in multimedia networks

The traffic modelling and its application to real traffic in operational networks, allows the implementation of research platforms that simulate future or real network critical conditions, which is particularly interesting for huge service providers. Injecting traffic series generated accordingly to mathematical models may help to evaluate several conditions in a network and certainly this may help to develop more accurate capacity planning models regarding specific QoS requirements. Such procedures also facilitate the creation of management strategies. A large number of tools on the Internet provide traffic analysis, like TG (TG), NetSpec (NetSpec), Netperf (Netperf), MGEN (MGEN) and D-ITG (D-ITG) and GTAR, Gerador de Tráfego e Analisador de QoS na Rede (Carvalho et al., 2006), FracLab (FracLab, 2011).

To model the traffic in integrated networks is necessary the use of mathematical models that allow, from its base, to infer the impact of traffic on network performance. The efficient characterization of traffic will be given by the degree of accuracy of the model in comparison with the real traffic statistical properties.

In our work, the characterization of the traffic is used as a key element in the design of complex telecommunications systems. Once characterized, the traffic on different time scales can be used in network simulations. The simulation process can reproduce the behaviour of traffic by application type, for parts of the network, by customer group or interconnections with other networks, opening the possibility to increase the knowledge of the network and making possible a better control of resources.

2.2 Poisson and erlang model

The use of the Internet to transmit real-time audio and video flows increases every day. Some of these applications are transmitted at a constant rate. This kind of traffic results by sending one packet every $1/Tx$ seconds, where Tx is the rate of transmission in packets per second, defined by the type of the application.

In circuit switched networks, a very successfully model is based on the Poisson distribution. The Poisson traffic is characterized by exponentially distributed random variables to represent the inter-packet times. The Erlang model, broadly used in telephony systems has been successfully used for capacity planning for many years and is based in the premise that a Poisson distribution describes the traffic in this type of network.

The Poisson model was considered accurate in the early years of the packet switched networks and was heavily used for capacity planning. In the early 90's, the work of Leland(Leland et al., 1994) proved that the behavior of the Ethernet traffic was considerably different than Poisson traffics mainly regarding self-similar aspects with long-range dependence, which is not well described by short memory processes. In practice, the packet switched networks that were planned using the Poisson model, normally had an overprovision in links capacity to comply with the lack of accuracy of the model. Considering the different works about capacity planning following the work of Leland, the heavy-tail models were considered more accurate to describe the traffic in packet switched networks and consequently, they appeared as a better choice.

2.3 Self-similar

One kind of traffic that appears often in wideband networks is the burst traffic. It can be generated by many applications such as compressed video services and file transfers. This traffic is characterized by periods with activity (on periods) and periods without activity (off periods). Moreover, as proved in (Perlingeiro & Ling, 2005), (Barreto, 2007), it is possible to generate self-similar traffic by the aggregation of many sources of burst traffics that presents a heavy-tailed distribution for the on period.

The self-similar model defines that a trace of traffic collected at a time scale has the same statistical characteristics that an appropriately scaled version of the traffic to a different time scale (Nichols et al., 1998). From the mathematical point of view, the self-similarity of a stochastic process in continuous time is defined as shown in Equation 1, which defines a process in continuous time X (t) as exactly self-similar.

$$X(t)\overset{d}{=}a^{-H}X(at), a>0 \tag{1}$$

The sample functions of a process $X(t)$ and its scaled version of the a-HX(at) obtained by compressing the time axis by the factor amplitudes "a" , can not be distinguished statistically. Therefore, the moments of order n of $X(t)$ are equal to the moments of order n of X (at), scaled by a-Hn. The Hurst parameter, H is then a key element to be identified in the traffic. For self-similar traffic, the H is greater than 0.5 and less than 1. For a Poisson traffic this value is close to 0.5. Experimental results show that this same parameter in operational networks (Perlingeiro & Ling, 2005; Carvalho et. Al., 2007) has values between 0.5 and 0.95. Then, the parameter H may be a descriptor of the degree of dependence on long traffic (Zhang et al.; 1997).

The aforementioned Hurst parameter plays a major role on the measurement of the self-similarity degree. The closer it is of the unity, the greatest the self-similarity degree. One of the most popular self-similar processes is the fractional Brownian motion (fBm), which is the only self-similar Gaussian process with stationary increments. The increments process of the fBm is the fractional Gaussian noise (fGn). To generate the traffic, we first create a fGn sequence based on the method presented in (Norros, 1995). Each sample of the sequence represents the number of packets to be sent on a time interval of size T. The size of the time interval and the mean of the sequence generated will depend on the traffic rate.

2.4 Multifractal traffic

As self-similar models, multifractals are multiscale process with rescaling properties, but with the main difference of being built on **multiplicative** schemes(Incite, 2011). In this way, they are highly non-Gaussian and are ruled by different limiting laws than the additive CLT (Central Limit Theorem). Therefore, multifractals can provide mathematical models to many world situations such as Internet traffic loads, web file requests, geo-physical data, images and many others. The Hölder function is defined by the h(t) function.

In the self similar model, also called as monofractal, the Hurst parameter is a global property that quantifies the process changes according to changes in the scale. For multifractal traffic, however, the Hurst parameter becomes less efficient in this characterization and another metric is needed to perform the scaling analysis of the sample regularity.

There are several ways to infer the scaling behavior of traffic, one way is widely used by local singularities of the function. A singular point is defined as a point in an equation, curve, surface, etc., which have transitions or becomes degenerate (Ried et al., 2000). It is quite common that the singular points of the signal containing essential information on network traffic packets.

In order to identify the singularities of a signal, it is necessary to measure the regularity of the same point, which will reflect in burst periods occurring at all traffic scales. In (Gilbert & Seuret, 2000) some examples can be found about the point and the exponents of the local Hölder values making possible to check the degree of uniqueness of network traffic.

According to Veira, (Veira et al., 2000) the Hölder exponent is capable to describe the degree of a singularity. Considering a function $f : R \rightarrow R$, with x_0 as real number, and α a stricted real positive number. It can be assumed that f belongs to $C_\alpha(x_0)$ if a polynomial P_m with degree $n < \alpha$, as shown in (2).

$$|f(x) - Pm(x - x_0)| \leq C|x - x_0|^\alpha \qquad (2)$$

As described in (Ludlam, 2004) a multifractal measure P can be characterized by calculating the distribution $f(\alpha)$, known as the multifractal, or singularity, spectrum where α is the local Hölder exponente (Clegg, 2005 ; Castro e Silva, 2004 ; Vieira, 2006). This measure can be also shown as a probability density function $P(x)$, in this case, the local Hölder exponente (; Gilbert & Seuret, 2000) is defined ad in (7).

$$\alpha(x) = \lim l \rightarrow 0 \log P(\mathcal{B}(l, x)) \log l \qquad (3)$$

where $\mathcal{B}(l, x)$ is a box centred at x with radius l, and $P(\mathcal{B})$ is the probability density integrated over the box $\mathcal{B}$. It describes the scaling of the probability within a box, centred on a point x , with the linear size of the box.

Each point x of the support of the measure will produce a different $\alpha(x)$, and the distribution of these exponents is what the singularity spectrum $f(\alpha)$ measures. The points for which the Hölder exponents are equal to some value α form a set, which is in turn a fractal object. The fractal dimension of this set can be calculated, and is a function of α, namely $f(\alpha)$.

As described in (2), a function $f(x)$ satisfies the Hölder condition in a neighborhood of a point, where c and n are constants, as in (4).

$$x_0 \text{ if } |f(x) - f(x_0)| \le c |(x-x_0)|^n \tag{4}$$

And a function $f(x)$ satisfies a Hölder condition in an interval or in a region of the plane, for all x and y in the interval or region, where c and n are constants, as in (5).

$$|f(x) - f(y)| \le c |x - y|^n \tag{5}$$

3. Traffic characterization

The process of traffic characterization is a preponderant point of a feasible network project. In this section a traffic characterization framework is described. The characterization intends to describe a step by step procedure, which may be useful to understand the behavior of traffic in large networks using a mathematical model as a tool to achieve good planning. One difficult issue to characterize traffic in IP networks is the changing environment due to new applications and new services that are appearing constantly. This implies that the

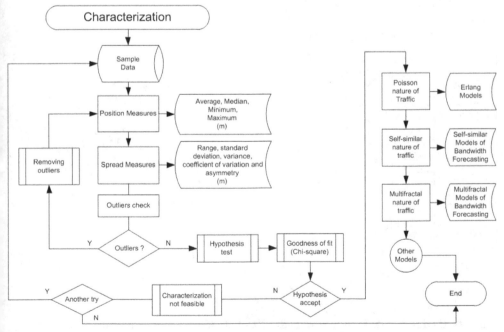

Fig. 2. Characterization process.

characterization used in real environments shall considerer the evolution and the amount of variation in the types of services, including not well known agents as social behavior and emerging applications.

The efficiency in traffic characterization is given by the model accuracy when compared with real traffic measures. As said by (Takine et al., 2004) a traffic model can only exist if there is a procedure for efficient and accurate inference for the parameters of the same mathematical structure. The traffic characterization is the main information source for the correct mathematical interpretation of network traffic. Once characterized, the traffic may be reproduced in different scales and periods and inserted into network simulators.

Figure 2 shows a complete characterization flow to optimize planning. This procedure was implemented in the GTAR (Barreto, 2007) simulator, developed within our research.

4. Experimental analysis

4.1 Analysis of an IP network

The first network to be evaluated is a Brazilian Service Provider in Brazil, with more than ten million PSTN (Public Switched Telephone Network) subscribers and more than one million ADSL as well. The IP network is shown Figure 3 each access layer is a PPPoX router capable called BRAS(Broadband Router Access Server).

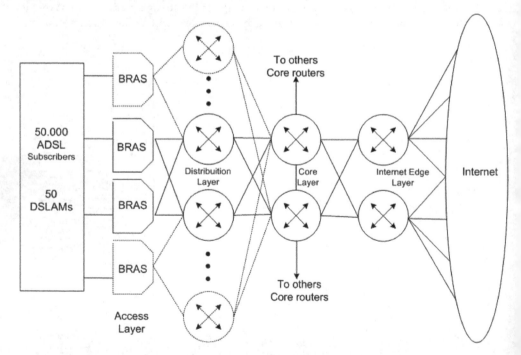

Fig. 3. Testbed Network Architecture with 40% of simultaneous attached subscribers at least, all IP/MPLS interface 1 or 10 Gigabit Ethernet, also for long distance. (De Deus, 2007).

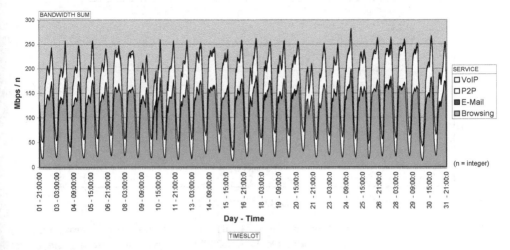

Fig. 4. Downstream traffic "on peak" and "off peak". The rate is normalized, 31 days sampled (De Deus, 2007).

Figure 4 and 5 shows the downstream traffic collection results for a 31 days period. The most important source of traffic is the HTTP(Browsing) following by P2P applications(e-Donkey, Bitorrent, Kazaa). In Figure 10, the same analysis is made for a 24 hours period.

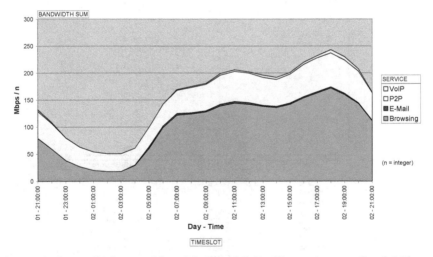

Fig. 5. Downstream traffic "on peak" and "off peak". Traffic rate is normalized, 24 hours sampled (De Deus, 2007).

Figure 6 shows the packet size probability distribution. Less than 100 Bytes packets have 50% of probability. These samples are from a real network with Internet traffic of 4 million xDSL subscribers, demonstrating the very large use of voice packets even when using http flows. This happens mainly because of applications such as SKYPE.

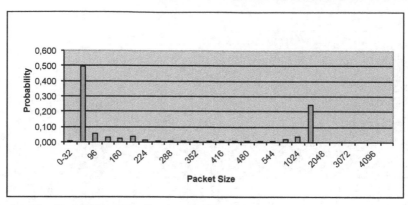

Fig. 6. Packet Size Probability Distribution (De Deus, 2007).

Table 1 shows and per application anaylis of traffic in which the Hurst parameter was calculated with two different methods (De Deus, 2007). For real time traffic, the Hurst parameter calculation demands attention because in some cases if statistical process does not have a representative long range dependence characteristic the parameter may be wrongly interpreted. Another issue is the trend present in the periodic traffic. For a more accurate estimation, the cycle regularity is removed to delete all observed trends.

Day	Hurst (Variance-Time Plot)	Hurst (Kettani-Gubner)	Chi-Square (Gaussian Distribution)
1	0,843	0,895	31,042
2	0,812	0,878	71,299
3	0,901	0,926	38,146
4	0,9	0,934	52,569
5	0,815	0,879	32,549
6	0,816	0,904	62,042
7	0,865	0,906	17,91
8	0,87	0,916	39,653
9	0,907	0,935	28,028
10	0,867	0,919	21,785
11	0,869	0,906	27,167
12	0,671	0,861	35,778
13	0,878	0,909	36,208
14	0,839	0,894	44,604
15	0,874	0,907	30,611
16	0,753	0,85	23,292
17	0,851	0,914	40,299

Day	Hurst (Variance-Time Plot)	Hurst (Kettani-Gubner)	Chi-Square (Gaussian Distribution)
1	0,915	0,948	63,549
2	0,942	0,962	49,771
3	0,937	0,963	45,25
4	0,902	0,935	28,243
5	0,902	0,928	20,708
6	0,901	0,942	30,181
7	0,939	0,964	43,258
8	0,932	0,964	52,354
9	0,937	0,968	39,007
10	0,922	0,948	38,576
11	0,86	0,926	37,5
12	0,904	0,942	33,84
13	0,937	0,965	55,799
14	0,922	0,958	46,972
15	0,935	0,963	46,757
16	0,935	0,967	49,986
17	0,933	0,964	37,285

Table 1. HTTP and P2P Hurst parameter estimation for 5 minutes average.

In Table 1 is shown the estimation of the H parameter for the HTTP (Hyper Text Transfer Protocol) applications. As can be seen, the H relies value between 0.67 and 0.93, which also shows a higher degree of self-similarity, considering that the lower value appears just in one day. For the P2P applications, the H parameter relies between 0.86 and 0.96.

The estimation of the the Hurst parameter in Table 1 uses three different methods: the Variance-Time Plot Method, the Kettani-Gubner Method (Clegg, 2005), (Barreto, 2007). Also a Chi-squared analysis was made as a non-parametric test of significance (Perlingeiro, 2006), (De Deus, 2007), (Clegg, 2005) due to the fact that it is necessary to verify the distribution similarity. The statistical significance test allows, with a certain degree of confidence, the acceptance or rejection of a hypothesis, as shown in Figure 7. The sampled links had a load, in the worst case around 70%.

Figure 7 shows the Hölder calculation for the traffic. The conclusion in fact is that the traffic is self-similar and monofractal, when the measurement is done in a 5 minutes per sample.

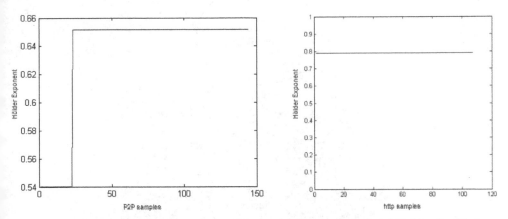

Fig. 7. P2P and http 5 minutes samples, Hölder exponent using the local Hölder Oscillation Based method [fraclab].

4.1.1 Bandwidth control strategies for the IP network

Figure 8 shows the proposal of a real-time network forecast. First, in the network the samples are collected. Then the traffic is classified per application. The estimation and a characterization of the parameters of collected samples are calculated. These parameters are used as input to a traffic forecast tool based on a mathematical traffic model which intends to find the sub-optimal capacity of the link for that traffic load, considering its self-similarity nature.

The objective is to use these parameters as inputs of a simulation tool to forecast the traffic and feedback in real-time the network to provide a new model to capacity plan in the backbone.

Following Figure 8, first the network samples are collected. Thenext step is the execution of classification procedure per application using tools based on protocols (Destination, Source, Port, Payload types). Next phase is to estimate the parameter (e.g. Hurst, Hölder) that will be used as input to a traffic forecast tool based on valid models (Norros *at al.*, 2000). The next step is to insert the parameter to a tool that will take a decision of how the auto-configuration will be done and a configuration of the element abstracting the vendor (e.g. Juniper, Cisco, Huawei). In figure 7 the example of application of the feedback process is described using the auto configuration tool to change the tunnel characteristics, that will use the proposed framework in Figure 8, as an example of setting up an outstream traffic marked as Diffserv.

If the traffic can be characterized as asymptotical self-similar or monofractal or multifractal some ready prediction models based (e.g. fBm, MWM, MMW) can be used. The core idea is that using only some parameters the mathematical calculus can be feasible at real time, as shown in Figure 14.

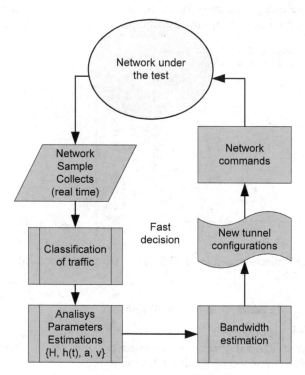

Fig. 8. Proposal of a network real-time forecast framework with bandwidth estimation.

In this case, the tunnels are configured using the self-similarity bandwidth estimators, as described in (Carvalho, 2007). The traffic needs to be marked as the DiffServ and will be injected per tunnel as the auto configuration tunnel selection.

There are several methods used to estimate bandwidth. The method used in our example is the FEP(Fractal Envelope Process). This model has a good performance for long range dependence with a high degree of confidence in the quasi-Real Time estimation (De Deus, 2007).

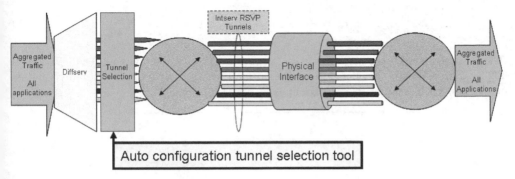

Fig. 9. Tunnel selection between two routers using Diffserv and Inteserv to select the specific tunnel.

The bandwidth estimation most accepted definition, currently known, use a concept introduced by (Kelly et al.,1996), where there is a direct dependency on buffer size and time scales related to the buffer overflow possibility. The concept is shown in (6) where X[0, t] is the amount of bits that arrive in an interval [0, t], considering that X[0, t] has stationary increments. The letter b is the buffer size and t time or scale, BP is the capacity in bits per second.

$$BP(b,t) = \frac{\log E\left[e^{bX[0,t]}\right]}{bt} \quad 0 < b, t < \infty \tag{6}$$

Based on this theory, several bandwidth estimators have been proposed and evaluated for its effectiveness and complexity of evaluation. In (Fonseca et al., 2005) an evaluation of the FEP estimator model (Fractal Envelope Process) was developed with good results, for use in high speed networks.

Equation (7) represents the FEP process estimation where the K is the buffer, a is the average, H is the Hurst parameter, σ is the standart devitation and Ploss represents the probability of packet loss when a buffer overflow. This is only valid when 0.5 < H < 1.

$$EN = \overline{a} + K^{\frac{H-1}{H}} * \left(\sqrt{-2*\ln(P_{loss})} * \sigma\right)^{\frac{1}{H}} * H(1-H)^{\frac{1-H}{H}} \tag{7}$$

Using (7) and correcting with (8) and (9), some curves are plotted in different time scales in Figure 10 (FEP Estimator and FEP Model). The best results are with 5 and 1 minutes, achieving the most next to average but still providing a good service with no delay, jitter or packet loss. The "Modelo FEP" f_{op} means the dynamic calculation per hour, the "Tunel P2P constante" and "Tunel HTTP constant"means the estimation with a Poisson Distribution Estimator, the P2P and HTTP means the dynamic bandwidth calculation.

$$f_{op} = \frac{2}{5}\frac{EN}{\sqrt{b'L}} \quad if \ 0.5 < H \leq 0.7 \tag{8}$$

$$f_{op} = \frac{2}{75} \frac{EN}{\sqrt{b'L}} \; if \; 0.7 < H < 1 \qquad (9)$$

The f_{op} is calculated based on (Perlingeiro & Ling, 2005) study as shown in (8) and (9), where EN is from (7) and b' is the normalized buffer ($b'=b/b_0$), where b is the buffer and b_0 minimum possible buffer size, L is the burst factor.

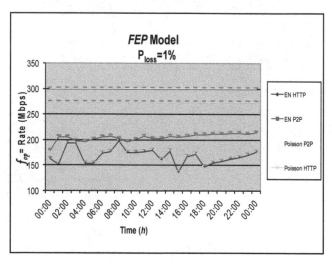

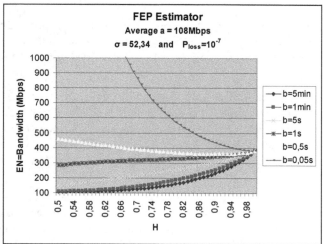

Fig. 10. Bandwidth estimation curves using the FEP method.

The FEP Model shown in Figure 10 uses a dynamic tunnel configurator as shown in Figure 9, denoting a better usage of the total available bandwidth. In the figures it appears that when a constant calculated bandwidth is used, more bandwidth is required. In the same way, the FEP Estimator shows that as much aggregated the traffic will be in any time scale, the difference will be minimum. In the other hand, when going to small time scales .05, .5 or 1 seconds, there is a trend in super estimation, proportional to the diminishing of the Hurst parameter.

As shown in many works (Leland et al., 1994), (Abry et al., 2002), (Carvalho et al., 2009), the Hurst parameter can show an accurate and single way to determinate the self-similarity. The Erlang model is very useful because its simplicity. A traffic engineer only needs to have some little information about service demand such as Retention time, blocking Probability, Number of Calls in the maximum usage hour to have the traffic and number of channels or resource needed.

The curves in Figure 10 show the possibility to have something, not so easy as Erlang model, but also possible to be achieved as a traffic model when a self-similar characterization is feasible. Also, the multi fractal model can also help to understand this same traffic in smaller scales, or in some case depending the traffic nature.

4.2 Analysis of a 3G network

The second evaluated network is a brazilian 3G network. This network runs with more than 1 million attached 3G costumers with national coverage. The traffic samples were collected in July, 2009 in three different locations (Leblon, Barra da Tijuca and Centro) in Rio de Janeiro. Two monitors were located in the network to collect the traffic, as shown in Figure 11.

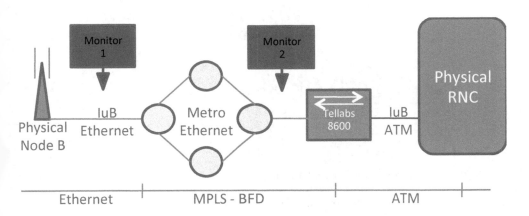

Fig. 11. 3G Network.

The main objective in this section is to investigate planning and project deployment phases based on traffic characterization. The first step is to classify the traffic per application. The second is to characterize the traffic using a procedure based on self-similarity (Clegg, 2005) or multifractal analysis (Carvalho et al, 2009).

Figure 12 shows the network topology for the Ethernet physical node B (ATM node) with an ATM-IP router which is responsible to convert ATM to Ethernet(IP). The same situation is found in RNC side where a Tellabs ATM-IP router aggregates all node B physical uplinks, every one carried through a Metro Ethernet network, with more than 50km radius Rio de Janeiro metropolitan area coverage.

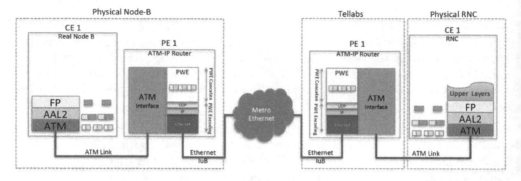

Fig. 12. 3G topology from Node B to RNC.

The first performance analysis of this network found some drawbacks in terms of latency and packet loss and jitter. In Figure 13 (before) is shown the first measures. One detected problem was the high level of broadcasting (ARP included) for this metro Ethernet network, in some periods, more than 80% of all IP traffic.

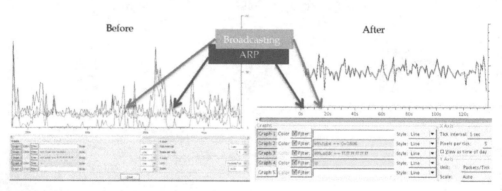

Fig. 13. 3G Traffic analysis (before and after).

As shown in Figure 12, the transport from Node-B to the RNC is performed by a MetroEthernet network that uses also a BFD protocol to track the availability of a Multiprotocol Label Switching (MPLS) Label Switched Path (LSP). In particular, BFD (Aggarwal et al., 2010) can be used to detect a data plane failure in the forwarding path of an MPLS LSP.

LSP Ping is an existing mechanism for detecting MPLS LSP data plane failures and for verifying the MPLS LSP data plane against the control plane, making possible the PseudoWire connections through a MPLS environment.

The problem, in this case, was an architectural design mistake because all Node B uplinks were configured in Level 2 VLANs (OSI Model), with more then 250+ 3G nodes B in the same IP subnet. The solution for this architectural problem was divide the Node Bs in 20 per subnet, as shown in Figure 8 (after).

This division resulted in diminishing the broadcasting to less than 5%. This problem is very simple in a typical Ethernet topology, but not so easy to be detected when inserted in a 3G network. Ethernet is a protocol designed for local area purposes; the MEF (Metro Ethernet Forum) inserted some signalling standards as a way to simplify the application in metro and long-range use.

Figure 14 shows the traffic trace collected in the 3G network and Figure 15 and Figure 16 show the singularity spectrum and the Hölder function for the 3G samples, showing the possibility to use the multifractal model also to forecast purposes.

This information is important do show this traffic can be characterized as multifractal in small scales of time, but in other hands the bandwidth model for this type of traffic model is also hard to build, because the nature of the traffic. Other important thing to understand is how to insert modifications with make the system not stable. In small scales, huge systems will need a lot of information to compute the bandwidth between to distance nodes.

The use of one model type can be very carefully choose because this could make the Operations Staff make wrong decisions that could result in many downtime.

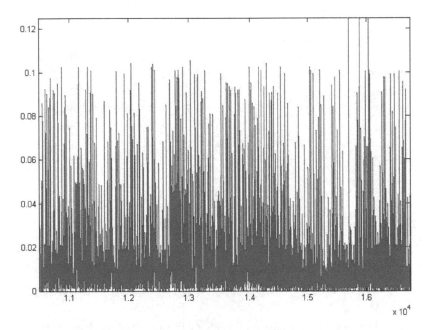

Fig. 14. Normalized 3G Traffic samples (milliseconds time scale).

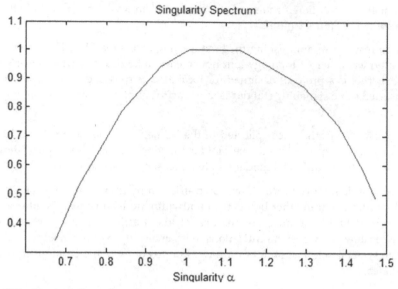

Fig. 15. 3G Traffic multifractal analysis – Singularity Spectrum (milliseconds time scale).

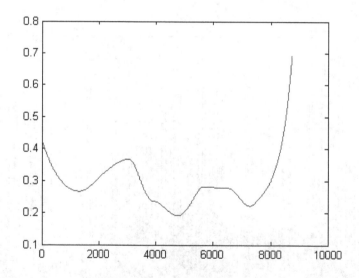

Fig. 16. 3G Traffic multifractal analysis – Hölder function. (milliseconds time scale).

5. Conclusion

This chapter presented an approach and a set of frameworks to characterize traffic and optimize network planning in IP and 3G networks. Based on real traffic measurements, we characterized the traffic and showed examples of how to apply the proposed frameworks. An special interest of our work has a focus in real operating networks and the examples show the application of the proposed frameworks in these environments.

The traffic characterization procedures for mutilmedia traffic were explained. We provided analyses by collecting different types of traffic and measuring its self-similar or multifractal degrees. All of this work was done with some self-developed (Carvalho et al., 2006) tools and also with some other tools (FRACLAB, 2011; OPNET, 2011).

The traffic models give us a good idea of the traffic behavior. In fact, the models can be valuable tools to the conception, management and sizing of a telecom network, resulting on efficient use of its resources. The operator can plan the growth of the network just to fit the business model, guaranteeing at different moments the efficient use of network resources, guaranteeing, on the other hand, the users satisfaction. In this context, the traffic models can also be used to define alternative policies that, for example, promote the network adaptation in periods with different levels of congestion.

Some very important to considering is how to improve the planning function with a better forecasting (Zukerman et al., 2003)., in terms of long time period for new assets plan and also to implement new products.

Something also very important is how to manage the network resources to have the best optimization possible, this will provide costumer better experience when using and buying exactly their needs.

6. References

Abry P., Baraniuk R.; Flandrin P.,Ried; R., Veitch D. (2002). The Multiscale Nature of Network Traffic Discovery, Analysis and Modeling. IEEE Signal Processing Magazine, 19(3):28-46.

Aggarwal, R; Kompella, K.; Nadeau, T.; Swallow, G. (2010). Bidirectional Forwarding Detection (BFD) for MPLS Label Switched Paths (LSPs). RFC 5884. ISSN: 2070-1721, June 2004. IETF Documents.

Avallone, S.; Pescapè, A.; Romano, S.P., Esposito, M.; Ventre, G. (2002). "Mtools: a one-way-delay and round-trip-time meter" 6th WSEAS International Conference, Crete, July 2002.

Barreto, P. S. (2007). Otimização de Roteamento Adaptativo em Redes Convergentes com tráfego autosimilar. Orientador:Carvalho, P. H. P. Tese de Doutorado, UnB.

Carvalho, P. H. P.; Barreto, P. S.; De Deus, M.; Queiroz, B.; Carneiro, B. (2007). A per Application Traffic Analysis in a Real IP/MPLS Service Provider Network. The 2nd IEEE IFIP/ International Workshop on Bradband Convergence Networks(IM2007/BCN2007), IEEE Communications Society , Munich, Germany, 21 a 25 de maio de 2007 ISBN: 1-4244-1297-8. Digital Object Identifier: 10.1109/BCN.2007.372751.

Carvalho, P.H.P.; Barreto, P. S.; Queiroz, B.; Carneiro, B.N. (2006). Modelagem, Geração e Análise de Tráfego em Redes Multiserviços, GTAR, LEMOM, UnB.

Carvalho, P.H.P.; Deus, M. A.; Barreto, P. S. (2009). Effective Bandwidth Allocation for IP/MPLS networks with Multimedia Traffic. In Portuguese : Alocação de Banda Efetiva para Tráfego Multimídia em Redes IP/MPLS. In: I2TS 2009, 2009, FLORIANOPOLIS. 8th International Information and Telecommunication Technologies Symposium, 2009, 2009.

Carvalho, P.H.P.; De Deus, M. A.; Barreto, P. S.; Fraga, T. ; Paiva, V. (2008). Identificação de Características Multifractais para Tráfego de Redes. In: XXVI Simpósio Brasileiro de Telecomunicações (SBrT'08), 2008, Rio de Janeiro, RJ. Anais do XXVI Simpósio Brasileiro de Telecomunicações, 2008.

Castro e Silva, J.L. (2004) "ProCon - Prognóstico de Congestionamento de Redes de Computadores usando Wavelets", Tese de Doutorado, Universidade Federal de Pernambuco, 2004.

Clegg, Richard (2005). A Practical Guide to Measuring the Hurst Parameter, Proceedings of 21st UK Performance Engineering Workshop, School of Computing Science.

D-ITG software (Sep 20, 2011) [Online]. Available:
http://www.grid.unina.it/software/ITG/index.php

De Deus, M.A (2007). IP/MPLS Bandwidth Management Strategies for Transport of Integrated Services. Estratégias de Gerenciamento de Banda IP/MPLS para os transporte de Serviços Integrados. Orientador : Carvalho, P.H.P ; Co-orientador : Barreto, P. S.; [Distrito Federal] 2007. xvii, 127p., 210 x 297 mm, ENE/FT/UnB, Mestre, Engenharia Elétrica, Comunicação(2007). Dissertação de Mestrado – Universidade de Brasília. Faculdade de Tecnologia.

Evans, J.; Filsfils, C. (2007). Deploying IP and MPLS QoS for Multiservice Networks. Morgan Kaufmann, ISBN-13: 978-0-12-370549-5.

Fonseca, N. L. S.; Drummond, A. C.; Devetsikiotis, M. (2005). Uma Avaliação de Estimadores de Banda Passante Baseados em Medições. Instituto de Computação–Universidade Estadual de Campinas. Department of Electrical and Computer Engineering – North Carolina State University Raleigh, USA

Fraclab (2011), A fractal analysis toolbox for signal and image processing. Available from http://fraclab.saclay.inria.fr/

Gilbert, A. e Seuret, S. (2000). Pointwise Hölder exponent estimation in data network traffic, In ITC Specialist Seminar, 2000.

Huang, J. (2000) "Generalizing 4IPP Traffic Model for IEEE 802.16.3", IEEE 802.16.3c-00/58, Meeting #11, Ottawa, Dec. 2000.

Incite (2011) Available: http://www.ece.rice.edu/INCITE/modeling_synopsis.html

Karagiannis, T; Faloutsos, M.; Riedi, R.H. (2002) "Long-Range Dependence: Now You See It, Now You Don't!" Proc. IEEE Global Telecommunications Conf. Global Internet Symposium, 2002.

Kelly, F.P.; Zachary, S.; Ziedins, I.; editors (1996). Notes on Effective Bandwidth, pages 141–168. Oxford University Press.

Kettani, H.; Gubner, J.A. (2002). "A Novel Approach the Estimation of the Hurst Parameter in Self-similar Traffic", Proceedings of IEEE Conference on Local Computer Networks, Tampa, Florida, November 2002.

Law, A.M; Kelton, W. D. (1991). "Simulation Modeling and Analysis", 2nd ed. New York: McGraw-Hill, 1991.

Leland, W. E; Taqq, M. S.; Willinger, W.; Wilson, D.V. (1994). On Self-similar nature of Ethernet traffic. ACM Sigcomm. Computer Communication.

Ledesma, S.; Liu, D. (2000). "A Fast Method for Generating Self-Similar Network Traffic", Proceedings of the 2000 International Conference on Communication Technologies, Beijing, China, p.54-61, Aug. 2000.

Ludlam, J. (2004). Localisation of the Vibrations of Amorphous Materials. PhD, Thesis Dissertation. Trinity College, Cambridge, UK, 2004. Online at: http://jon.recoil.org/thesis/thesisse11.xml

Melo, E. T. L. (2001). "Qualidade de Serviço em Redes IP com DiffServ: Avaliação através de Medições", 2001.

MGEN software (Sep 20, 2011) [Online]. Available: http://mgen.pf.itd.nrl.navy.mil/

Netspec software (Aug 28, 2011) [Online]. Available: http://www.ittc.ku.edu/netspec/

Netperf software (Jun 7, 2011) [Online]. Available:
http://www.netperf.org/netperf/NetperfPage.html

Norros, Ilkka. (1995). On the use of factional Brownian motion in the theory of connectionless networks. IEEE Journal of Selected Areas in Communications, 13(6):953-962.

Opnet (2011), http://www.opnet.com.

Paxson, V. (2000) "Fast, approximate synthesis of fractional Gaussian noise for generating self-similar network traffic", Computer Communication Review, vol.27, p.5-18.

Perlingeiro, F. R.; Ling, L. L.. (2005). Estudo de Estimação de Banda Efetiva para Tráfego auto-similar com variância infinita, SBrT'05, 04-08 de setembro de 2005, Campinas, SP

Riedi, R. H. ; Ribeiro, V. J. ; Crouse, M. S. and Baraniu, R. G. (2000). Network Traffic Modeling Using a Multifractal Wavelet Model. Proceedings European Congress of Mathematics, Barcelona 2000. Department of Electrical and Computer Engineering, Rice University, 6100 South Main Street Houston, TX 77005, USA (NSF/DARPA).

Takine, T.; Okazaki, K.; Masuyama, H. (2004). IP Traffic Modeling: Most Relevant Time-Scale and Local Poisson Property. Department of Applied Mathematics and Physics Kyoto University. (ICKS'04) Informatics Research for Development of Knowledge Society Infrastructure

Taqqu, M. S.; Willnger, M.S.W.; Sherman, B. (1997). "Proof of a fundamental result in self-similar traffic modeling". Computer Communication Review, vol. 27, p. 5-23, 1997.

TG software (Aug 8, 2011) [Online]. Available: http://www.postel.org/tg/

Vieira, F.H.; Jorge, C.; e Ling, L. (2005) Predição Adaptativa do Expoente de Hölder para Tráfego Multifractal de Redes, In XXVIII Congresso Nacional de Matemática Aplicada e Computacional, 2005.

Vieira, Flavio H. T. V. (2006). Contribuições ao cálculo de banda e probabilidade de perda para tráfego multifractal. Tese de Doutorado. Unicamp, 2006.

Zhang, H.F.; Shu, Y.T.; Yang, O. (1997). Estimation of Hurst parameter by variance-time plots. Communications, Computers and Signal Processing, 1997. apos;10 Years

PACRIM 1987-1997 - Networking the Pacific Rimapos;. 1997 IEEE Pacific Rim Conference on Volume 2, Issue , 20-22 Aug 1997 Page(s):883 - 886 vol.2

Zukerman, M.; Neame, T. D.; Addie, R. G. (2003). "Internet Traffic Modeling and Future Technology Implications" Proceedings of Infocom, 2003.

Part 2

Quality of Services

4

A Testbed About Priority-Based Dynamic Connection Profiles in QoS Wireless Multimedia Networks

A. Toppan, P. Toppan, C. De Castro and O. Andrisano

IEIIT-CNR, National Research Council of Italy & WiLab,
University of Bologna, Bologna,
Italy

1. Introduction

The ever-growing demand of high-quality broadband connectivity in mobile scenarios, as well as the Digital Divide discrimination, are boosting the development of more and more efficient wireless technologies.

Despite their adaptability and relative small installation costs, wireless networks still lack a full bandwidth availability and are also subject to interference problems.

In context of a Metropolitan Area Network serving a large number of users, a bandwidth increase can turn out to be neither feasible nor justified. In consequence, and in order to meet the needs of multimedia applications, bandwidth optimization techniques were designed and developed, such as Traffic Shaping [1-3], Policy-Based Traffic Management [4-8] and Quality of Service (QoS) [9-17].

In this paper, QoS protocols are adopted and, in particular, priority-based dynamic profiles in a QoS wireless multimedia network. This technique [18-20] allows to asssign different priorities to distinct applications, so as to rearrange service quality in a dynamic way [21,22] and guarantee the desired performance to a given data flow.

In particular, the platform can manage two levels of priority: among different users and within a single user's connection.

In the former kind of priority management, those users whose guaranteed bandwidth is higher, will be proportionally assigned a greater part of the shared bandwidth.

The latter case refers to each single user, whose distinct services are assigned distinct priorities. Each profile, in fact, allows the real time management of services, and the priority parameter is used to queue the desidered services properly.

A complete testbed involving 80 users approximately is here presented, where such technique is adapted to the specific requirements of the plant.

The network infrastructure installation is detailed, the whole QoS system developed is described and four measurement campaigns are reported.

The whole testing was directed by WiLab (www.wilab.org), which includes the IEIIT-CNR (National Research Council of Italy, IEIIT Bologna unit) and a portion of the TLC scientific community at Bologna University (Italy). The design and technical aspects of the problem were and are still being carried out by such group.

The proposed platform aims at supporting a Wireless Internet Service Provider (WISP) in the management of its network infrastructure in a user-friendly and straightforward way. It can be accessed through the Internet and lets the network manager define different access profiles and supervise all the users' connections.

The QoS service, in particular, allows to set each user's minimum bit rate guaranteed and maximum supported, enable services such as VoIP, FTP, Mail and P2P and assign them specific priorities.

The network scenario installed and used for the testbed includes an Internet gateway, a server which hosts the whole infrastructure control system and five sectoral distribution devices.

The software platform allows to define some distinct kinds of priority-based connection profiles, each characterized by a set of different parameters and a diverse commercial value. Personal data can also be managed, each corresponding to the user's kind of connection profile subscribed.

Reports about connections and traffic statistics are also at disposal, also useful to law purposes. A continuous monitoring of the wireless network infrastucture is also possible.

All the above features can be easily managed through specific user-friendly portals.

This kind of services are fundamental in many application fields, ranging from Intelligent Transportation Systems (ITS) and Infomobility to "Smart Cities", where wireless applications guide the user in most of his activities.

The paper is organized as follows: Section 2 describes the testbed setup and the QoS software developed. Section 3 details the results of the four measurements campaigns.

These techniques, addressed to to real applications, are discussed in the following: the PEGASUS project about the support of real time in Infomobility is discussed in Section 4. The Smart Cities scenario is presented in Section 5. Conclusions and future tesbed extensions are discussed in Section 6.

2. The network infrastructure and QoS system

Although the main purpose of this work is to present measurements, it is important to describe the testbed setup and some related installation problems, as well as the QoS system and its main principles.

2.1 Testbed setup

The network plant is depicted in Fig. 1 and, from the left to the right, involves a Shelter for Internet distribution (A) which, due to property reasons, could not host the QoS Management Server (B), installed 1.2 km away (Link 1).

Such system runs on a Dual Xeon 2 GHz, 8Gb di Ram, 80Gb SAS Raid 5 server and comprehends a Radius authentication server, a PPPoE concentrator and the whole QoS management software.

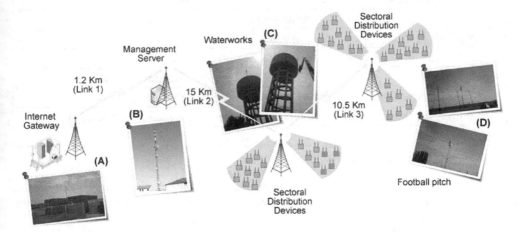

Fig. 1. the QoS equipment.

A HiperLAN link starts here and connects the management server to a waterworks area (C) 15 km away (Link 2). In such area, two sectoral distribution equipments were setup and serve 35 users approximately.

A further antenna allows to reach the last distribution area, situated near a football pitch (D) 10.5 km away (Link 3), and including three further sectoral distribution equipments for 45 users approximately. Such antenna had to be setup since some unfavourable features of the ground prevented a direct connection from being created.

The indirect link creates a bottleneck and forces the waterworks area to support some of the traffic surrounding the pitch.

A further penalty is that the same pylon which hosts the antenna in area (C) also carries some television and microwave aerials. In consequence, some devices not optimally shielded were initially blocked and even damaged and the available bit-rate is still being diminished.

In addition, some further installation problems were caused by the daily activation of the waterworks pump, which produced strong perturbations to the electric network, consequent blocks of many devices and even breakdowns. This problem was solved by means of an electronic filter.

All the above problems would have obviously prevented 80 users from being propely served, unless a bandwidth optimization method and traffic management were adopted.

2.2 The QoS architecture adopted

The QoS scheme is based on the dynamic assignment and redistribution of bandwidth on the basis of priority and users' profiles. The main parameters of each kind of profile are summed up in Tab. 1. In particular, when QoS management is enabled in a profile (QoS flag), different priorities can be assigned to distinct protocols.

Parameter	Description
Name	Profile name
Description	Profile description
Upload bandwidth	Max upload bandwidth (kbit/s)
Upload guaranteed	Min upload bandwidth guaranteed (kbit/s)
Download bandwidth	Max download bandwidth (kbit/s)
Download guaranteed	Min download bandwidth guaranteed (kbit/s)
QoS	Flag for enabling QoS traffic management

Table 1. main fields of a connection profile.

Fig. 2 shows the graphic interface for the definition and management of each type of profile. The seven panels "Band 1",. .. "Band 7" on the right allow to assign priorities, 1 being the highest, 7 the lowest. In particular, a protocol can be associated to a specif priority by dragging and dropping its name in the chosen band panel.

The minimum bandwidth guaranteed (in percentage) and maximum available must also be setup for each sub-bandwidth. Once a profile has been defined, it can be assigned to many distinct users, so as to tailor service supply easily and quickly.

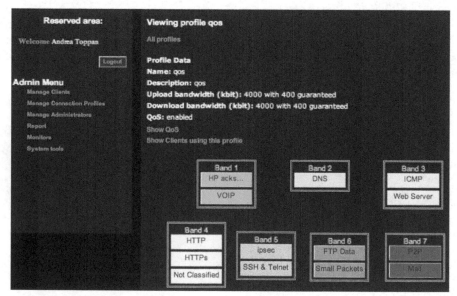

Fig. 2. priority assignment through drag&drop operations.

As far as dynamic QoS management is concerned, the basic idea is to limit both upload and download operations through the Egress policier (www.egress.com), so as to discard all the packets whose speed exceeds a maximum value set. To this purpose, queuing algorithms are applied and bandwidth can consequently be tuned according to actual availability.

Some problems had to be solved along the way: shaping could only be applied to the outgoing traffic, already processed by the kernel, whereas both the uploading and downloading flows should normally be shaped by queuing methods on both the incoming and outcoming interfaces. In this case, though, many independent PPP interfaces were simultaneously active and each PPPoE was thus identified through a PP system interface numbered N (PPN, N = 0, 1,. ..).

In consequence, two further difficulties had to be faced:

1. too many iptables rules were generated and so was a further branch in the queuing structures;
2. the htp qdisc bandwidth sharing capabilities could not be fully exploited and no minimal bandwidth per PPP connection could be guaranteed.

As a matter of fact, each PPP having its own independent queuing, the traffic on the network interface was managed in an unpredictable way: no minimum bandwidth per connection could be even assigned and the unexploited bandwidth could not be dynamically and equally redistributed among all connections.

In order to handle the above situation, a qdisc (common to all connections) and subclasses for each kind of connection (with minimum and maximum bandwidth set) were defined, so as to redistribute unexploited resources among tunnels.

To this purpose, a hierarchical structure based on HTB (Hierarchical Token Bucket, http://luxik.cdi.cz/~devik/qos/htb) queuing was developed, whose nodes specify their own minimum and maximum bandwidth. In this way, the traffic of each tunnel is forwarded to the class it pertains to, so as to achieve the desired result.

Nevertheless, qdiscs can only manage the traffic of their own interface, so it was still impossible to identify a single connection by accessing the network interface of the PPPoE server. Each connection, in fact, is managed as a separate network interface.

An IMQ (Intermediate Queueing Device, www.linuximq.net) interface was thus adopted, which allows to manage qdiscs and the whole traffic: iptables are deviated to such interface and traffic can be shaped. Each single PPP interface must be assigned a connection identifier and sent to the IMQ, where connection classes were defined.

In this way, each connection can monitored, traffic can be classified on the basis of protocols and the most important flows are assigned the highest priority.

Another problem faced was that each packet could only be marked by means of an an identifier, so, theoretically speaking, the simultaneous identification of connections and protocols within a session was impossible.

Several tests demonstrated that the problem could be solved through the joint use of u32filter+MARK and CLASSIFY TARGET. This was done defining a further HTB class structure in the PPPN interfaces.

The bandwidth redistribution problem having been solved, attention could be focused on flow priority within each connection.

Fig. 3 shows how queuing algorithms were applied. A root node (qdisc) was created in the imq0 interface; class 1:1 was added in order to define the total bandwidth (100 Mbit/s in this case). Subclasses were then defined for the management of single connections, each specifying the minimum guaranteed (rate) and maximum at disposal (ceil).

Note that a higher QoS could be achieved if the SFQ (Stochastic Fair Queuing) were applied, so as to manage single flows through a Round Robin policy.

In order to manage priority of single flows, a hierarchic structure was created within each PPPN interface.

As in the former case, several subclasses were added to the root node (qdisc), each with a minimum and maximum bandwidth; the SFQ algorithm was applied afterwards. A u32filter list is defined in the root node of each interface, so as to drive single packets on the class they pertain to on the basis of their protocol. Packets were initially divided by the CLASSIFY TARGET tool according to the connection; a further division was then made by u32filter+MARK on the basis of protocols.

Besides, the identifiers range is 1:10000 and 1:65535 in hex, so the highest attention must be paid to each class handles. The following synatx was adopted:

iptables -t mangle -A POSTROUTING -j CLASSIFY --set-class x:y

In this way, filters could be avoided and everything is managed by iptables.

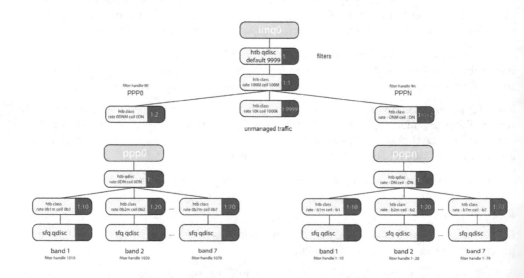

Fig. 3. hierarchical QoS management.

2.2.1 Statistics

The platform allows to visualize information about the infrastructure and its use and, in consequence, to make statistics about connected users and their traffic volume.

Fig. 4, for instance, refers to all the users' total connection time. For the sake of compactness, only a small excerpt was reported.

Diagrams are also available describing the system components, such as CPU load, network load and many others.

Reserved area:	Global statistics (Users online in the last 24 hours)			
Welcome Andrea Toppan				
[Logout]	User_id	Connection time	Upload	Download	Online
Admin Menu		161 days, 4 hours, 33 minutes, 10 seconds	78.69 GB	112.18 GB	ONLINE
Manage Clients		12 days, 20 hours, 47 minutes, 6 seconds	19.51 MB	208.73 MB	OFFLINE
Manage Connection Profiles		189 days, 11 hours, 59 minutes, 20 seconds	17.96 GB	33.32 GB	ONLINE
Manage Administrators		230 days, 23 hours, 25 seconds	0.81 GB	6.06 GB	ONLINE
Report		105 days, 1 hours, 10 minutes, 13 seconds	130.13 MB	1.33 GB	ONLINE
Monitors		187 days, 16 hours, 6 minutes, 24 seconds	51.65 GB	78.91 GB	ONLINE
		104 days, 12 hours, 12 minutes, 25 seconds	3.81 GB	12.07 GB	ONLINE
		191 days, 9 hours, 59 minutes, 53 seconds	2.31 GB	6.88 GB	ONLINE
		123 days, 5 hours, 12 minutes, 23 seconds	18.01 GB	23.85 GB	ONLINE
		13 days, 23 hours, 30 minutes, 18 seconds	31.34 MB	283.81 MB	ONLINE
		250 days, 22 hours, 13 minutes, 34 seconds	3.00 GB	34.96 GB	ONLINE
		96 days, 7 hours, 42 minutes, 36 seconds	9.28 GB	23.43 GB	ONLINE
		200 days, 18 hours, 22 minutes, 15 seconds	12.37 GB	76.01 GB	ONLINE
		95 days, 19 hours, 23 seconds	2.07 GB	8.91 GB	OFFLINE

Fig. 4. small excerpt from all the users' total connection time.

As far as each user is concerned (Fig. 5), the following data can be monitored: connections, data volumes exchange, diagrams about his or her traffic and, as indicated by law, packets logging.The authentication system adopted is Radius and traffic is encapsulated through the PPPoE protocol. In consequence, a PPP tunnel is active between each user and the server.

3. Measurements campaigns

As already anticipated, the main concern of this paper is to present a realistic testbed for QoS management; four measurements campaigns carried out are described in the following, whose scenario was described in Section 2.1 (Fig. 1).

The adopted dynamic priority-based method is twofold. On the one hand, users are assigned the shared bandwidth on the basis of their profiles: the higher the guaranteed bandwidth, the higher the shared bandwidth assigned. On the other hand, not only users but also services within a profile are prioritized, so each user is aware that his or her bandwidth is accordingly shared among his or her applications.

The QoS management is presented in an increased way: no QoS in the first campaign; QoS with neither minimum nor maximum set in the second; minimum and maximum bandwidth defined in the third; dynamic redistribution is also managed in the fourth.

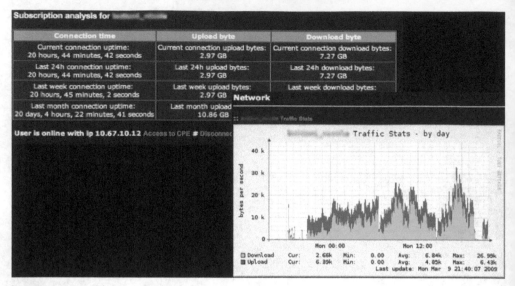

Fig. 5. statistics about a single user.

3.1 Throughput on single links (no QoS)

The first campaign represents throughput maximization in the single links of the whole infrastructure. Fig. 6 shows results in Link 1, from the Shelter to the Management Server.

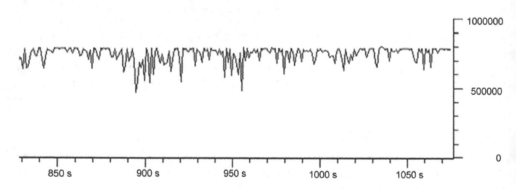

Fig. 6. throughput in the link from the Shelter to the Management Server (bit/s).

The above measurements were performed using the following tools: (1) Iperf (www.noc.ucf.edu/Tools/Iperf), which allows to send a TCP or UDP data streams and measure their throughput; (2) WireShark (www.wireshark.org), a network analyzer which allows to capture and diagram Iperf streams.

In this campaign, the Iperf server was installed on an Acer Travelmate with Linux Debian OS and located in the B node, so as to receive UDP connections. A Macbook Pro with Mac OSX Leopard was used as the client.

Nodes A, C, D hosted laptops for the connection to the Iperf server. In this way, throughput could be measured first in Link1, then Link2 and Link 3 and finally in Link2 + Link3.

The highest UDP throughput at disposal in Iperf connnections was set to 10 Mbit/s; results are reported in Tab. 2 and derive from a large number of measurements properly mediated.

Note that the throughput from the Management Server to the football pitch (Link 2 + Link 3) is almost 2Mbit/s lower than in single links 2 and 3: this derives from the same device being charged of both signal reception and transmission.

Link	Description	Throughput
Link 1	From the Shelter to the Management Server	7.8 Mbit/s
Link 2	Management Server to Waterworks	6.6 Mbit/s
Link 3	Waterworks to Football pitch	6.9 Mbit/s
Link2 + Link 3	Management Server to Football pitch	5.2 Mbit/s

Table 2. results in single links.

3.2 QoS applied to multimedia TCP flows (no min/max bandwidth set)

The second campaign aimed at verifying the efficiency of the QoS management server. The Iperf was moved in the (C) node, so as to check Link 3.

An ethernet cable subsituted the wireless connection during a PPP connection. Delays and packet loss, in fact, are not particularly relevant in this kind of control, attention being mainly focused on flow management.

The following tools were adopted: (1) QoS Server; (2) Server-side Vlc for MMS over http video flow transmission (www.videolan.org/vlc); (3) Client-side Vlc for flow reception; (4) WireShark.

Fig. 7, diagrammed through WireShark, represents the scenario and the first measurements of this campaign. It refers to the following profile: no QoS applied, symmetric upload and dowmlod of 1Mbit/s, no bandwidth guaranteed. The client receives the first video (red line) until second 230, then the second video (green line) starts and the firts one is interrupted at second 280.

As expected, in the concurrent period (sec. 230 to 280) both TCP videos are blocked, the bandwidth being inadequate to support both of them.

Fig. 8 represents the second test and involves three videos on three distinct ports; in this case, a QoS profile was enabled which guarantees an increasing priority from the first to the third flow.

First starts the video on the lowest priority port (blue curve); the intermediate priority video starts 20 seconds later (green curve) and, in consequence, the first data flow declines. In the period between sec. 40 to 80 the maximum throughput was increased, so as to emphasize the effect of dark and still scenes in the second video. In this way, the total throughput is constant and bandwidth waste is kept under control.

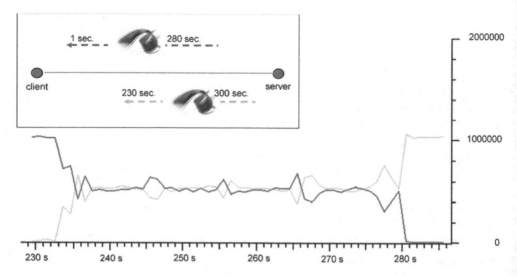

Fig. 7. two concurrent video flows without QoS (bit/s): scenario and results.

The same observations apply to the third and highest priority flow (red curve): the three videos share the bandwidth and, thanks to QoS, the lowest priority one is flattened, the medium is assigned less bandwidth and the highest has the best quality.

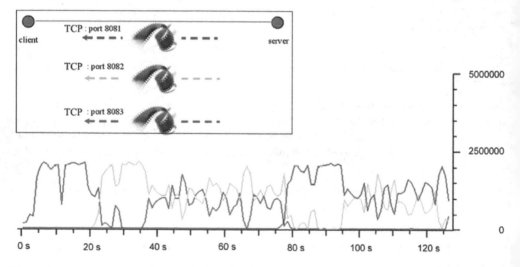

Fig. 8. three concurrent videos with QoS (bit/s).

3.3 Bandwidth control in QoS management

As the previous measurements showed, the QoS policy adopted helps to avoid bandwidth waste and guarantees a better service, especially for VoIP, IPTV support etc.

Nevertheless, this kind of priority management among traffic classes implies the almost complete cancellation of the less important services for the benefit of the most important ones.

In the initial QoS profile, an increasing priority was assigned to the three TCP flows on distinct ports. Fig. 9 shows the measures obtained. The less priority flow (blue curve) is strongly limited by the second one (green curve). They are both flattened when the highest priority flow starts (red curve).

In order to improve such results, each traffic class was then assigned a minimum and a maximum throughput (Tab. 3). The third campaign tries to demonstrate the effectiveness of such method and Fig. 10 reports the results.

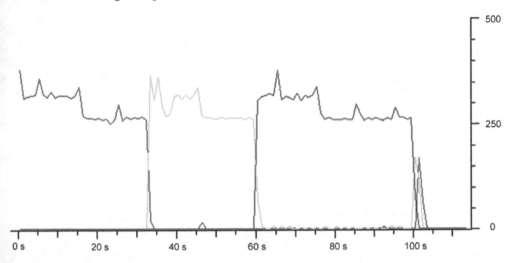

Fig. 9. TCP concurrent flows with QoS (bit/s).

Class	Priority level	Min. bandwidth % guaranteed	Max. bandwidth % at disposal
1	Highest priority	0%	100%
2	Intermediate	30%	30%
3	Lowest priority	20%	100%

Table 3. minimum throughput guaranteed and maximum available, as assigned to single flows.

Note that an upper bound having been imposed to the intermediate flow, the lowest is not totally flattened, but only slowed down.

The most priority flow starts at second 50 and, in consequence, the less priority traffic becomes slower, but not more than the 20% guaranteed. The same applies to the intermediate flow, for which a 30% at least is available. The highest priority flow, of course, can not reach the maximum speed.

As this measure shows, if guaranteed bandwidth percentages are properly managed, a high QoS can be obtained in an easy and immediate way.

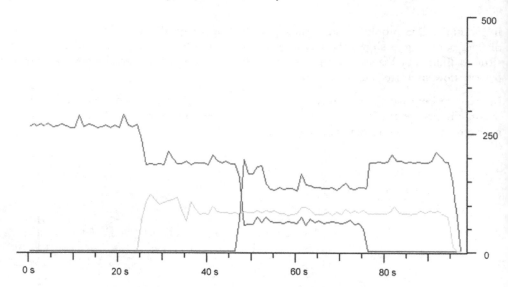

Fig. 10. TCP concurrent flows with bounded bandwidth QoS (bit/s).

3.4 Proportional reassignment of bandwidth

An important feature that must be handled in this kind of QoS management is bandwidth reassignment proportionally to each user's minimum guaranteed.

In this case, three PCs and the usual tools were adopted and two kinds of profiles were defined (Tab. 4).

Clients	Min. guaranteed	Max. at disposal
Clients 1 and 2	128 Kbit/s	3 Mbit/s
Client 3	600 Kbit/s	3 Mbit/s

Table 4. Minimum and maximum throughput for each client.

Clients were connected to the server through the PPPoE protocol; the maximum throughput between clients and server was set to 3Mbit/s, so as to simulate a set of wireless relays.

In this case, the upper bandwidth was to be shared among concurrent users and, in consequence, none was to reach the maximum.

Results are shown in Fig. 11: initially, the only connected client 1 (green curve) gets the whole bandwidth available according to his profile (3 Mbit/s).

After 100 seconds approximately, client 2 is also connected, throughput is assigned according to the minimum guaranteed and exceeding bandwitdh is reassigned according to such value.

The two clients share the same profile, so the bandwidth is equally divided and redistributed.

At second 150 the third client (red curve) connects to the system; the minimum throughput of his or her profile is four times the others', so clients 1 and 2 are limited accordingly.

When client 1 disconnects, bandwidth is distributed among clients 2 and 3 in a ratio of 1 to 4.

Finally, client 3 logs out and the whole bandwidth is at client's 2 disposal.

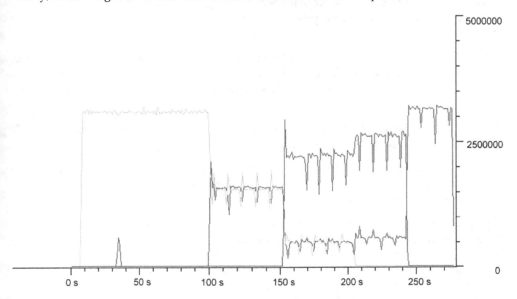

Fig. 11. bandwidth proportional reassignment among connected users.

4. The PEGASUS project: Real time support in infomobility services

One of the most important applications to which QoS techniques have been applied are Smart Navigation and the braoader field of Infomobility.

Transportation is one of the main fields where advanced technological systems can improve human life in a significant way: risks due to accidents, time wasted travelling and pollution could be highly reduced by applications for vehicle localization, behaviour prediction, etc.

These considerations are at the basis of the increasing interest that ITS are gaining in these years.

Furthermore, the latest study on global urbanization conducted by the Population Division of the Department of Economic and Social Affairs of the United Nations predicts that, in 2050, nearly 70% of the global population will be living in larger cities [23].

This immense aggregation of people will surely pose great challenges to the sustainability of modern lifestyle, and the problem of an efficient management of mobility stands out as one of the most relevant ones.

As a matter of fact, densely populated cities imply the concentration (from the country) and distribution (within the city) of massive amounts of people and resources [24].

In addition to the vast economic importance and consequences of such situation, urban and sub-urban mobility is a serious challenge also due to the circulation of large amounts of people and goods in a relatively small area. This poses hazards to life and health, especially for children, the elderly, and unfamiliar visitors, as well as to the environment.

Urban mobility, in fact, accounts for some 30% of energy consumption and 70% of transport pollution in Europe, and this problem is magnified by the increasing population concentration in large cities.

In such a scenario, the efficient management of traffic is a challenge that governments, industries and researchers are forced to face worldwide. Private travellers, commercial road users, and the public sector are continually searching for new and faster travel routes and methods.

Roads efficiency can be substantially improved by the deployment of ITS, which exploit ICT in order to provide traffic safety and efficiency.

ICT can be considered as the foundation for carrying out smart navigation, meant as the paradigm where mobile entities (vehicles and pedestrians) move wisely through a given environment, exploiting reliable and timely information about traffic conditions.

In this context, one of the most important applications is the support of real time, meant as the constant monitoring of traffic and road conditions, and the consequent possible update of the routes previously suggested. As a matter of fact, the best path in a given situation can vary when traffic conditions vary and updates should be notified to the user in real time. Nevertheless, up to now, no simple and marketable product was proposed for monitoring traffic and providing real time information to road users.

To this purpose, in the framework of the Italian project PEGASUS (http://pegasus.octotelematics.com/), WiLab aims at exploiting information transmitted from vehicles to a remote Control Center, so as to provide drivers in real time with updated information about actual traffic conditions. In this way, a new smart navigation service is supported. In particular, the objective is twofold:

- investigate the impact of smart navigation on the communication networks load;
- investigate the impact of real time updates on traffic management efficiency; as a matter of fact, vehicles equipped with smart navigators are constantly sent information about actual roads conditions;

In Fig. 12, the smart navigation scenario considered and developed at WiLab is shown: vehicles are equipped with on-board units (OBUs), which periodically transmit their speed and position (known through the GPS integrated on board) to a Control Center. Such data are transferred through the General Packet Radio Service (GPRS) network.

The fleet equipped with OBUs is addressed as *floating car data (FCD)*. In March 2010, the Italian FCD to which the PEGASUS project refers, reached over 1.000.000 equipped vehicles (OctoTelematics, 2010); this number is to increase quickly (note that the number of public and private vehicles in Italy was 34 million back to 2003 [25], hence the FCD is a not negligible percentage of the overall private vehicles number).

All such data are processed and exploited for the real time dynamic navigation of vehicles (hereafter Dynamic Route Guidance, DRG); the same information can also be forwarded to public or private institutions for traffic management, etc.

In the near future, almost all vehicles will be able to send real time information, and the majority of drivers will take profit of data properly processed and of applications beyond traffic management, such as safety and entertainment.

In this scenario, telecommunications systems will be required to transmit information quickly and reliably, both among vehicles and between vehicles and remote control centers. Which technologies are to be chosen, how priorities must be managed, which capacity is required, are still open issues.

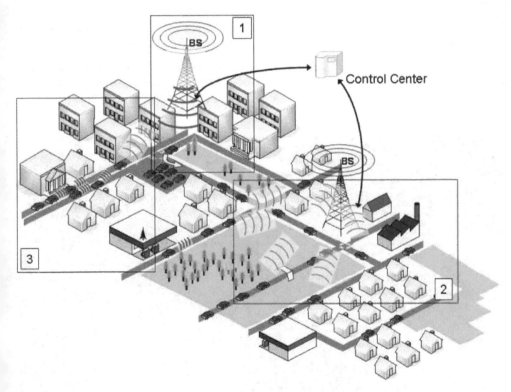

Fig. 12. Smart navigation scenario.

The mobile network is, at present, the only one adopted for vehicles-Control Center communication; nevertheless, the quantity and size of information is to increase.

Urban networks, thus, can turn out to be a precious support for existing infrastructures, especially if properly managed through effective QoS techniques.

5. The smart cities scenario

The QoS testbed is currently being applied to Smart Cities scenarios, a class of applications which are gaining an increasing attention.

As described by William J. Mitchell (MIT, Smart Cities Group, http://cities.media.mit.edu/): "Our cities are fast transforming into artificial ecosystems of interconnected, interdependent intelligent digital organisms. Emerging applications in the ICT field are poised to reshape our urban environments".

In this context, wireless architectures and QoS infrastructures become nodes of a TLC network, which collects information from the surrounding areas and consequently supplies citizens with advanced infomobility services.

A very diffused and challenging problem to face is where hardware can be installed, since many areas and city centers are protected and regulated by severe rules by the Ministry of Culture and Heritage.

A possible solution is to place the whole hardware within preexistent structures, such as electricity posts, properly adapted in order to host the required technology.

A testbed based on this approach adopts "Intelligent Posts" hosting both hardware and software (Fig. 13). Such posts were installed by the WiLab group, in cooperation with Fondazione Almamater and Ghisamestieri within Villa Gandolfi Pallavicini (Bologna, Italy).

Fig. 13. Posts by Ghisamestieri for Smart Cities.

In this case, the effective management of QoS is tested in order to supply citizens with several services, such as video surveillance (both wired and wireless data transmission involved), integrated image analysis, Internet connection within urban environments, RFID services for tourists, emergency calls, radiodiffusion, fire prevention, parking management, localization, diagnostics and control by television. In addition, sensor network data collection, traffic information, access control, mobile payments, vehicle tracking, user-generated contents, energy management, etc.

Through QoS management, such services will be configured dynamically on the basis of bandwidth availability. According to the throughput actually available in a specific temporal slot and thanks to a constant monitoring of radio resources on each route, both services to be offered to the user and applications to be kept active can be chosen.

More specifically, the testbed is addressed to transport improvement and traffic reduction through smart navigators.

6. Conclusions and future QoS testbed extensions

Coming back to the tests in Sections 2, 3, PPP tunnels between the server and users can be temporarily closed, when packet transimission is slowed down by interference or machinery stops.

The first problem is that, in case the PPP LCP surveys trouble situations, the channel is closed and the client disconnected. An authomatic procedure is in charge of reconnection, but a time waste in PPP tunnel setup as well as abrupt disconnections are bound to take place.

A second problem derives from traffic limitation and control being handled by a single QoS server: in this case, data are properly limited only after they have crossed one or two links. In other words, in case an authenticated user sends an UDP data flow larger than his or her maximum upload bandwidth, such flow will be diminished only after reaching the QoS server. Meanwhile, the available bandwidth will be unproperly occupied by such flow.

On the basis of such considerations, the testbed will be extended according to two different scenarios of distributed QoS architectures [26-28]. The first one is depicted in Fig. 14 and aims at avoiding tunnel closure in case of interference and packet loss.

In case many relays occur between the client and PPPoE concentrator, packet loss can increase; the idea, thus, is to shorten the tunnel, so as to integrate the PPPoE concentrator and the transmitter. In this way, all PPP features could be maintained and its limitations diminished. The tunnel, in fact, would be established between the client's CPE and the nearest transmitter and communication between the transmitter and the main server could be based on TCP/IP.

Furthermore, if interferences between transmitter and main server would take place, packets could be relayed without PPP tunnel drops.

A disadvantage could concern uncoded communication between the main server and pylons. Possible solutions could be the activation of encrypted systems or a PPP tunnel to the main server. In this case, the user would not even perceive any link failure.

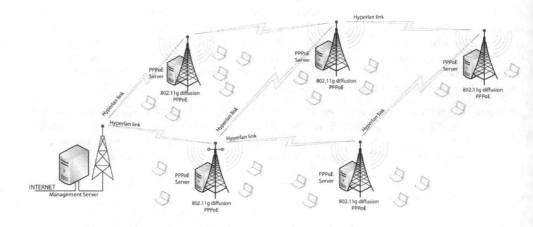

Fig. 14. first extended testbed scenario.

The second scenario (Fig. 15) aims at solving the second problem arisen: the idea is to apply the first control on users' bandwidth at the pylon.

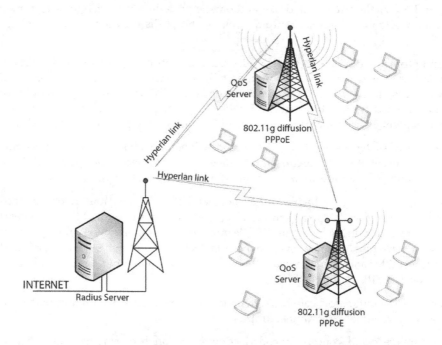

Fig. 15. second extended testbed scenario.

As for the PPPoE concentrator, the QoS manager itself could be integrated in the transmitter. On the one hand, this kind of control logic decentralization would solve the problem of link saturation in case of heavy UDP uploads. On the other hand, the server would be spared from an exceeding traffic in case of network expansion.

Two further difficulties arise: firstly, connection plans are not anymore managed by a single server in a centralized and transparent way. In consequence, a new communication protocol is required for the authomatic configuration of devices on the pylon when the main server configuration changes.

In addition, logic decentralization can cause more frequent failures of important components. If a failure occurs of QoS or PPPoE components, thus, an infrastructure is needed that prevents connections from being denied.

A switch, for instance, could be used to disconnect out of order devices and the main server would be in charge of guaranteeing connectivity until the problem is solved.

At present, the idea is to avoid a complete implementation of the above scenarios, using simulation tools instead. In particular (Fig. 16), the QoS impact could be evaluated through the joint use of the actually developed parts and simulators.

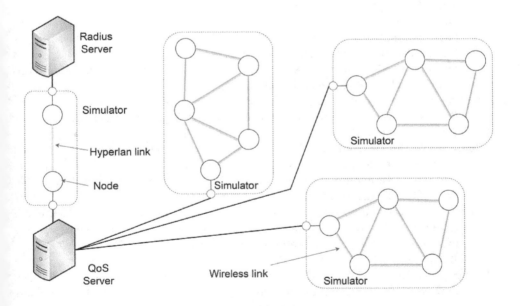

Fig. 16. A schema for the simulation of QoS server impact.

7. Acknowledgment

More than an acknowledgment, a dedication: To the little Gabriele Toppan, the son of Paolo and the nephew of Andrea, go our very best wishes to grow up strong, responsible and enthusiastic about life and its numerous miracles.

8. References

[1] Gringeri, S.; Shuaib, K.; Egorov, R. et al.; Traffic shaping, bandwidth allocation, and quality assessment for MPEG video distribution over broadband networks, Network, IEEE, 12 , 6, pp. 94 – 107, 1998, doi: 10.1109/65.752648

[2] Frank Yong Li; Stol, N.; QoS provisioning using traffic shaping and policing in 3rd-generation wireless networks, in Proc. of IEEE Wireless Communications and Networking Conference, 2002, 1, 139 – 143, 2002, doi: 10.1109/WCNC.2002.993478

[3] Yongdong Wang; Jurczyk, M.; Impact of traffic shaping in ATM networks on video quality, in Proc. of International Workshops on Parallel Processing, 1, 485 – 492, 2000, doi: 10.1109/ICPPW.2000.869154

[4] Gozalvez, D.; Monserrat, J.F.; Calabuig, D., et al; Policy-based channel access mechanism selection for QoS provision in IEEE 802.11e, Vehicular Technology Magazine, IEEE, 2, 3, 29-34, 2007, doi: 10.1109/MVT.2008.915326

[5] Flegkas, P.; Trimintzios, P.; Pavlou, G.; A policy-based quality of service management system for IP DiffServ networks, Network, IEEE, 16, 2, 50-56, 2002, doi: 10.1109/65.993223

[6] Fangming Zhao; Lingge Jiang; Chen He; Policy-based radio resource allocation for wireless mobile networks, in Proc. of IEEE International Conference on Neural Networks and Signal Processing, 476-481, 2008, doi: 10.1109/ICNNSP.2008.4590396

[7] Conchon, E.; Pérennou, T.; Garcia, J. Et al.; W-NINE: A Two-Stage Emulation Platform for Mobile and Wireless Systems, EURASIP Journal on Wireless Communications and Networking, 2010, Article ID 149075

[8] Heithecker, S.; do Carmo Lucas, A.; Ernst, R.; A High-End Real time Digital Film Processing Reconfigurable Platform, EURASIP Journal on Embedded Systems, 2007, Article ID 85318

[9] Chang Wook Ahn; Ramakrishna, R.S.; QoS provisioning dynamic connection-admission control for multimedia wireless networks using a Hopfield neural network, Vehicular Technology, IEEE Transactions on, 53, 1, 106-117, 2004, doi:10.1109/TVT.2003.822000

[10] Xiang Chen; Bin Li; Yuguang Fang; A dynamic multiple-threshold bandwidth reservation (DMTBR) scheme for QoS provisioning in multimedia wireless networks, Wireless Communications, IEEE Transactions on, 4, 2, 583-592, 2005, doi: 10.1109/TWC.2004.843053

[11] Huan Chen; Kumar, S.; Kuo, C.J.; Dynamic call admission control scheme for QoS priority handoff in multimedia cellular systems, in Proc. of IEEE Wireless Communications and Networking Conference, 114-118, 2002, doi: 10.1109/WCNC.2002.993474

[12] Yaya Wei; Chuang Lin; Fengyuan Ren et al; Dynamic priority handoff scheme in differentiated QoS wireless multimedia networks, in Proc. of Eighth IEEE International Symposium on Computers and Communication, 131-136, 2003, doi: 10.1109/ISCC.2003.1214112

[13] Xiaorong Li; Chuah, E.; Jo Yew Tham; Kwong Huang Goh; An optimal smooth QoS adaptation strategy for QoS differentiated scalable media streaming, in Proc. of IEEE International Conference on Multimedia and Expo, 429-432, 2008, doi: 10.1109/ICME.2008.4607463

[14] Huang, J.-H.; Li-Chun Wang; Chung-Ju Chang; Capacity and QoS for a Scalable Ring-Based Wireless Mesh Network, IEEE Journal on Selected Areas in Communications, 24, 11, 2070-2080, 2006, doi: 10.1109/JSAC.2006.881622

[15] Bai, B; Chen, W.; Cao, Z. et al; Uplink Cross-Layer Scheduling with Differential QoS Requirements in OFDMA Systems, EURASIP Journal on Wireless Communications and Networking, 2010, Article ID 168357

[16] Montazeri, S.; Fathy, M.; Berangi, R.; An Adaptive Fair-Distributed Scheduling Algorithm to Guarantee QoS for Both VBR and CBR Video Traffics on IEEE 802.11e WLANs, EURASIP Journal on Advances in Signal Processing, 2008, Article ID 264790

[17] Almeida, M.; Sarrô, R.; Barraca, J.P. et al; Experimental Evaluation of the Usage of Ad Hoc Networks as Stubs for Multiservice Networks, EURASIP Journal on Wireless Communications and Networking, 2007, Article ID 62967

[18] Ganesh Babu, T.V.J.; Le-Ngoc, T.; Hayes, J.F.; Performance of a priority-based dynamic capacity allocation scheme for wireless ATM systems, IEEE Journal on Selected Areas in Communications, 19, 2, 355-369, 2001, doi: 10.1109/49.914513

[19] Naser, H.; Mouftah, H.T.; A class-of-service oriented packet scheduling (COPS) algorithm for EPON-based access networks, in Proc. of 7th Int. Conference on Transparent Optical Networks, 232-236, 2005, doi: 10.1109/ICTON.2005.1505793

[20] Song, S.; Manikopoulos, C.N.; A Priority-based Feedback Flow Control System for Bandwidth Control, in Proc. of 40th Annual Conference on Information Sciences and Systems, 1645-1652, 2006, doi: 10.1109/CISS.2006.286399

[21] Zhang, F.; Verma, P.K.; Cheng, S.; Pricing, resource allocation and quality of service in multi-class networks with competitive market model, Communications, IET, 5, 1, 51-60, doi: 10.1049/iet-com.2009.0694

[22] Kamosny, D.; Novotyny, V.; Balik, M.; Bandwidth Redistribution Algorithm for Single Source Multicast Networking, in Proc. of Int. Conference on Systems and Int. Conference on Mobile Communications and Learning Technologies, 147-156, 2006, doi: 10.1109/ICNICONSMCL.2006.62

[23] UN, World urbanization prospects: The 2007 revision population database, 2008, http://esa.un.org/unup/

[24] EU, Eu mobility and transport, 2010, http://ec.europa.eu/transport/publications/statistics/

[25] ecoage, Independent ecology portal, 2003, www.ecoage.net

[26] Won-Kyu Hong, D., Choong Seon Hong, C.; A QoS management framework for distributed multimedia systems, Int. J. Network Mgmt, 13, 115-127, 2003, doi: 10.1002/nem.465

[27] Jing Li; Yongwang Zhao; Min Liu et al; An adaptive heuristic approach for distributed QoS-based service composition, in Proc. of IEEE Symposium on Computers and Communications (ISCC), 687-694, 2010, doi: 10.1109/ISCC.2010.5546721

[28] Pattara-Atikom, W.; Krishnamurthy, P.; Banerjee, S.; Comparison of distributed fair QoS mechanisms in wireless LANs, in Proc. of IEEE Global Telecommunications Conference GLOBECOM '03. 553-557, 2003, doi: 10.1109/GLOCOM.2003.1258298

5

End to End Quality of Service in UMTS Systems

Wei Zhuang
China Telecom Co. Ltd. (Shanghai)
P.R.China

1. Introduction

About ten years ago, WCDMA [1] based the third generation mobile systems started to be deployed worldwide. Besides to support basic mobile data services such as file transfer and internet surfer, etc., UMTS has one of the most significant archievements which can support a richer variety of services with QoS guarantee, such as video, VOIP, etc. Quality of Service (QoS) is defined as "the collective effect of service" performance, which determines the degree of satisfaction of a user of the service in the ITU-T recommendation E.800. At a technical level, QoS can be characterized by service availability, delay, jitter, throughput, packet loss rate (Nortel White Paper, 2002).

3GPP has put many efforts to define and standardize a QoS framework for data services, specially IP-based services. The standardization of a UMTS QoS model started in 1999. the development was based on the following key principles: operation and QoS provisioning needed to be possible in the wireless environment, usage of the Internet QoS mechanisms, applications and interoperability. This chapter is aimed to provide an overview of the UMTS end-to-end QoS architecture, describe how the QoS requirements to be realized from top layer to wireless links.

2. WCDMA QoS architecture

QoS standardization in UMTS PS domain enables UMTS to provide data service with end-to-end QoS guarantees. 3GPP proposed a layered architecture for supporting end-to-end QoS. It includes the following key elements (Sudhir Dixit et al., 2001):

- Mapping of end-to-end services provided by the UE, UTRAN, Core Network (CN), and external IP networks;
- Traffic classes and associated QoS parameters;
- Location of QoS functions;
- QoS negotiation;
- Multiplexing of flows onto network resources;
- An end-to-end data delivery model.

---

[1] Wideband Code Division Multiple Access W-CDMA - the radio technology of UMTS - is a part of the ITU IMT-2000 family of 3G Standards.

The layered UMTS QoS architecture is shown in Figure 1. The UMTS network can provide end-to-end QoS services from a Terminal Equipment (TE) to another TE. A network bearer service describes how to realize a certain network QoS. It is defined by the control signaling, user traffic transport and QoS management functionality, which enabling the provision of a contracted QoS (Sudhir Dixit et al., 2001; 3GPP23107, 2011). As the end-to-end service is conveyed over several networks, the end-to-end bearer service consists of different network bearer services. The end-to-end bearer service can be decomposed into TE/MT local bearer service, the UMTS bearer service and the external bearer service.

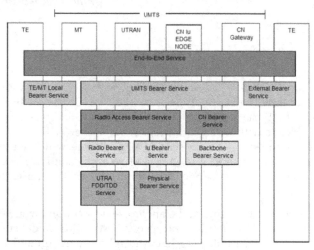

Fig. 1. UMTS QoS architecture

The TE/MT local bearer service provides communication between the TE and MT parts. MT (Mobil Terminal) provides connection to the UTRAN with basic functions, such as radio attachment to 3G network, authenticating the CS/PS domain, mobility management, etc. TE support call control, authenticating the IMS subscription, etc.

The external bearer service deals with the interoperability and interworking aspects with external IP bearer, and provides the appropriate functionality to support it. It is logical located in the GGSN, which is the gateway of UMTS to external network (Sotiris et al., 2002).

UMTS bearer service provides service by using the radio access bearer service (RAB) and the core network bearer service. The detail is given in the following.

2.1 UMTS bear service

The UMTS QoS is provided by the UMTS bearer service. It includes the radio access bearer service and the core network bearer service. They reflect the optimized way to realize the UMTS Bearer Service over the respective cellular network topology taking into account aspects such as mobility and mobile subscriber profiles.

The Radio Access Bearer Service provides confidential transport of signalling and user data between MT and CN Iu Edge Node with the QoS adequate to the negotiated UMTS Bearer Service or with the default QoS for signalling. This service is based on the characteristics of the radio interface and is maintained for a moving MT.

The Core Network Bearer Service of the UMTS core network connects the UMTS CN Iu Edge Node with the CN Gateway to the external network. The role of this service is to efficiently control and utilise the backbone network in order to provide the contracted UMTS bearer service. The UMTS packet core network shall support different backbone bearer services for variety of QoS. And the UMTS bearer service is realized by a GPRS service in the PS domain or a speech/data service in the CS domain.

2.1.1 The radio bearer service and Iu bearer service

The Radio Access Bearer Service is realised by a Radio Bearer Service and an Iu-Bearer Service. The role of the Radio Bearer Service is to cover all the aspects of the radio interface transport. This bearer service uses the UTRA FDD/TDD, which is not elaborated further in this chapter.

The Iu-Bearer Service together with the Physical Bearer Service provides the transport between UTRAN and CN. Iu bearer services for packet traffic shall provide different bearer services for variety of QoS.

2.1.2 The backbone network service

The Core Network Bearer Service uses a generic Backbone Network Service. The Backbone Network Service covers the layer 1/Layer2 functionality and is selected according to operator's choice in order to fulfill the QoS requirements of the Core Network Bearer Service. The Backbone Network Service is not specific to UMTS but may reuse an existing standard.

2.2 QoS requirement

2.2.1 UMTS QoS classes

The layered UMTS QoS architecture requires the definition of QoS attributes for each bearer service. When defining the UMTS QoS classes, the restrictions and limitations of the radio interface have to be taken into account. The QoS mechanism should be simpler than that in wired network due to different error characteristics of the air interface. Table 1 illustrates the QoS classes defined by 3GPP.

The main distinguishing factor between these QoS classes is how delay sensitive the traffic is.

Conversational class

The transfer time of real time conversation scheme shall be low because of the conversational nature of the scheme and at the same time that the time relation (variation) between information entities of the stream shall be preserved in the same way as for real time streams. The maximum transfer delay is given by the human perception of video and audio conversation. Therefore the limit for acceptable transfer delay is very strict, as failure to provide low enough transfer delay will result in unacceptable lack of quality. The transfer delay requirement is therefore both significantly lower and more stringent than the round trip delay of the interactive traffic case. The fundamental characteristic for QoS is to preserve time relation (variation) between information entities of stream and conversational pattern (stringent and low delay) (3GPP23107, 2011).

The most well known use of this scheme is telephony speech (e.g. GSM). But with Internet and multimedia a number of new applications will require this scheme, for example voice over IP and video conferencing tools. Real time conversation is always performed between

Traffic class	Conversational class conversational RT	Streaming class streaming RT	Interactive class Interactive best effort	Background Background best effort
Fundamental characteristics	- Preserve time relation (variation) between information entities of the stream Conversational pattern (stringent and low delay)	- Preserve time relation (variation) between information entities of the stream	- Request response pattern - Preserve payload content	- Destination is not expecting the data within a certain time - Preserve payload content
Example of the application	voice	streaming video	Web browsing	background download of emails

Table 1. UMTS QoS classes

peers (or groups) of live (human) end-users. This is the only scheme where the required characteristics are strictly given by human perception.

Streaming class

Streaming class is characterised by that the time relations (variation) between information entities (i.e. samples, packets) within a flow shall be preserved, although it does not have any requirement on low transfer delay (3GPP23107, 2011). This scheme is one of the newcomers in data communication, raising a number of new requirements in both telecommunication and data communication systems. It is a one-way transport. A user can use this class to watch (listen to) real time video (audio).

The delay variation of the end-to-end flow shall be limited, to preserve the time relation (variation) between information entities of the stream. But as the stream normally is time aligned at the receiving end (in the user equipment), the highest acceptable delay variation over the transmission media is given by the capability of the time alignment function of the application. Acceptable delay variation is thus much greater than the delay variation given by the limits of human perception. The fundamental characteristics for streaming class QoS is to preserve time relation (variation) between information entities of the stream.

Interactive class

Interactive traffic is the other classical data communication scheme that on an overall level is characterized by the request response pattern of the end-user. At the message destination there is an entity expecting the message (response) within a certain time. Round trip delay time is therefore one of the key attributes. Another characteristic is that the content of the packets shall be transparently transferred (with low bit error rate). The fundamental characteristics for interactive class QoS is to request response pattern, preserve payload content.

This class is applied when an end-user (either a machine or a human) is using on line requesting data from remote equipment (e.g. a server). Examples of human interaction with the remote equipment are: web browsing, data base retrieval, server access. Examples of machines interaction with remote equipment are: polling for measurement records and automatic data base enquiries (tele-machines).

Background class

Background traffic is one of the classical data communication schemes that on an overall level is characterised by that the destination is not expecting the data within a certain time. The scheme is thus more or less delivery time insensitive. Another characteristic is that the content of the packets shall be transparently transferred (with low bit error rate). The fundamental characteristics for background class QoS are a) destination is not expecting the data within a certain time; b) preserve payload content.

When the end-user, that typically is a computer, sends and receives data-files in the background, this scheme applies. Examples are background delivery of E-mails, SMS, download of databases and reception of measurement records.

2.2.2 UMTS bearer service attributes

UMTS bearer service attributes describe the service provided by the UMTS network to the user of the UMTS bearer service. A set of QoS attributes (QoS profile) specifies this service. At UMTS bearer service establishment or modification different QoS profiles have to be taken into account.

Traffic class ('conversational', 'streaming', 'interactive', 'background')

Traffic class is a type of application for which the UMTS bearer service is optimized. By including the traffic class itself as an attribute, UMTS can make assumptions about the traffic source and optimise the transport for that traffic type.

Maximum bit-rate (kbps)

Maximum bit-rate (kbps) is the maximum number of bits delivered to UMTS at a SAP (Service Access Point) within a period of time, divided by the duration of the period. The traffic is conformant with Maximum bit-rate as long as it follows a token bucket algorithm where token rate equals Maximum bit-rate and bucket size equals Maximum SDU size.

The algorithm is well known as "Token Bucket Algorithm" which has been described in IETF. It is a reference algorithm for the conformance definition of bitrate. This may be used for traffic contract between UMTS bearers and external network/user equipment. In the algorithm, "tokens" represents the allowed data volume, for example in byte. "Tokens" are given at a constant "token rate" by a traffic contract, are stored temporarily in a "token bucket", and are consumed by accepting the packet. This algorithm uses the following two parameters (r and b) for the traffic contract and one variable (TBC) for the internal usage.

- r: token rate, (corresponds to the monitored Maximum bitrate/Guaranteed bitrate).
- b: bucket size, (the upper bound of TBC, corresponds to bounded burst size).
- TBC (Token bucket counter): the number of given/remained tokens at any time.

According to a token bucket, conformance can be defined as: "Data is conformant if the amount of data submitted during any arbitrarily chosen time period T does not exceed (b+rT)".

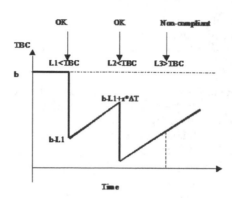

Fig. 2. Operation example of the reference conformance algorithm

The algorithm is described here (Figure 2, (3GPP23107, 2011)). Token bucket counter (TBC) is usually increased by "r" in each small time unit. However, TBC has upper bound "b" and the value of TBC shall never exceed "b". When a packet i with length Li arrives, the receiver checks the current TBC. If the TBC value is equal to or larger than Li, the packet arrival is judged compliant, i.e., the traffic is conformant. At this moment tokens corresponding to the packet length is consumed, and TBC value decreases by Li. When a packet j with length Lj arrives, if TBC is less than Lj, the packet arrival is non-compliant, i.e., the traffic is not conformant. In this case, the value of TBC is not updated.

The Maximum bitrate is the upper limit a user or application can accept or provide. All UMTS bearer service attributes may be fulfilled for traffic up to the Maximum bitrate depending on the network conditions. The downlink of the radio interface can use maximum bitrate to make code reservations. Its purpose is:

1. to limit the delivered bitrate to applications or external networks with such limitations;
2. to allow maximum wanted user bitrate to be defined for applications able to operate with different rates (e.g. applications with adapting codecs).

Guaranteed bitrate (kbps)

Guaranteed bitrate (kbps) is defined as: a guaranteed number of bits delivered by UMTS at a SAP within a period of time (provided that there is data to deliver), divided by the duration of the period. The traffic is conformant with the guaranteed bitrate as long as it follows a token bucket algorithm where token rate equals Guaranteed bitrate and bucket size equals Maximum SDU size.

UMTS bearer service attributes, e.g. delay and reliability attributes, are guaranteed for traffic up to the Guaranteed bitrate. For the traffic exceeding the Guaranteed bitrate the UMTS bearer service attributes are not guaranteed. Guaranteed bitrate may be used to facilitate admission control based on available resources, and for resource allocation within UMTS.

Delivery order (y/n)

Delivery order indicates whether the UMTS bearer shall provide in-sequence SDU delivery or not. This attribute is derived from the user protocol (PDP type) and specifies if out-of-sequence SDUs are acceptable or not. Whether out-of-sequence SDUs are dropped or re-ordered depends on the specified reliability.

Maximum SDU size (octets)

Maximum SDU size (octets) means the maximum allowed SDU size. The maximum SDU size is used for admission control and policing.

SDU format information (bits)

SDU format information (bits) is a list of possible exact sizes of SDUs. UTRAN needs SDU size information to operate in transparent RLC protocol mode, which is beneficial to spectral efficiency and delay when RLC re-transmission is not used. Thus, if the application can specify SDU sizes, the bearer is less expensive.

SDU error ratio

SDU error ratio indicates the fraction of SDUs lost or detected as erroneous. SDU error ratio is defined only for conforming traffic. It is used to configure the protocols, algorithms and error detection schemes, primarily within UTRAN.

Residual bit error ratio Residual bit error ratio indicates the undetected bit error ratio in the delivered SDUs. If no error detection is requested, Residual bit error ratio indicates the bit error ratio in the delivered SDUs. It is used to configure radio interface protocols, algorithms and error detection coding.

Delivery of erroneous SDUs (y/n/-)

Delivery of erroneous SDU (yes/no/-) indicates whether SDUs detected as erroneous shall be delivered or discarded. 'Yes' implies that error detection is employed and that erroneous SDUs are delivered together with an error indication, 'no' implies that error detection is employed and that erroneous SDUs are discarded, and '-' implies that SDUs are delivered without considering error detection. It is used to decide whether error detection is needed and whether frames with detected errors shall be forwarded or not.

Transfer delay (ms)

Transfer delay (ms) indicates maximum delay for 95th percentile of the distribution of delay for all delivered SDUs during the lifetime of a bearer service, where delay for an SDU is defined as the time from a request to transfer an SDU at one SAP to its delivery at the other SAP. Transfer delay can be used to specify the delay tolerated by the application. It allows UTRAN to set transport formats and ARQ parameters.

Traffic handling priority

Traffic handling priority specifies the relative importance for handling of all SDUs belonging to the UMTS bearer compared to the SDUs of other bearers. Within the interactive class, there is a definite need to differentiate between bearer qualities. This is handled by using the traffic handling priority attribute, to allow UMTS to schedule traffic accordingly. By definition, priority is an alternative to absolute guarantees, and thus these two attribute types cannot be used together for a single bearer.

Allocation/Retention priority

Allocation/Retention priority specifies the relative importance compared to other UMTS bearers for allocation and retention of the UMTS bearer. The Allocation/Retention

Priority attribute is a subscription attribute which is not negotiated from the mobile terminal. Priority is used for differentiating between bearers when performing allocation and retention of a bearer. Where there is no enough resource, the relevant network elements can use the Allocation/Retention Priority to prioritize bearers with a high Allocation/Retention Priority over bearers with a low Allocation/Retention Priority when performing admission control.

Source statistics descriptor ('speech'/'unknown')

Source statistics descriptor ('speech'/'unknow') specifies characteristics of the source of submitted SDUs. Conversational speech has a well-known statistical behaviour (or the discontinuous transmission (DTX) factor). By using source statistics descriptor, a network element can know whether the SDUs of a UMTS bearer generated by a speech source or not. UTRAN, the SGSN and the GGSN and also the UE may, based on experience, calculate a statistical multiplex gain for use in admission control on the relevant interfaces.

2.3 QoS management for UMTS bearer service

In the section, an overview of QoS functions is described which is used to establish, modify, and maintain a UMTS bearer service with a specific QoS. The allocation of these functions to the UMTS entities indicates the requirement for specific entity to enforce the QoS commitments negotiated for the UMTS bearer service. UMTS is split into user plane and control plane for easy expanding in the future. So QoS management functions are also split into user plane and control plane. All of the QoS management functions in both planes (control and user plane) will ensure the provision of the negotiated service between the access points of the UMTS bearer service. The end-to-end service is provided by translation/mapping with UMTS external services.

2.3.1 QoS management in control plane

The QoS management functions in control plane are shown in Figure 3. The QoS functions for UMTS bearer service include service manager, translation function, admission/capability control and subscription control in the control plane. These functions are used to establish and modify a UMTS bearer service through signaling/negotiating with UMTS external services, establishing/modifying UMTS internal services.

Subscription control checks the administrative rights when an UMTS bearer service user requires a service with the specified QoS. It is located at CN EDGE.

Admission capability control will maintains available resource information of a network entity, and resource allocated to UTMS bearer service. When receiving an UMTS bearer service request or modification request, admission / capability control function determines whether the required resources can be provided or not. If the network can provide resources, it will reserve these resources. This function also checks the capability of a network entity, i.e. whether the specific service is implemented and not blocked for administrative reasons. It is located at MT, UTRAN, CN EDGE and Gateway.

Translation function is used to convert between internal service primitives and external protocols. It is located at MT and Gateway. At MT and Gateway, translation function converts between UMTS bearer service attributes and external network QoS attributes.

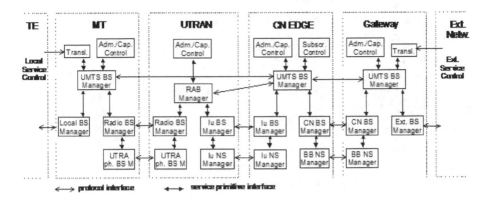

Fig. 3. QoS management function for UMTS bearer service in the control plane

Service manager co-ordinates the related functions in control plane to establish, modify and maintain the service. All user plane QoS management functions are supported by service manager with the relevant attributes. The service manager may perform an attribute translation to request lower layer services. Service manager at UMTS bearer service level is located at MT, CN EDGE and Gateway. The UMTS BS manager can signal among each other and via the translation function with external instances to establish / modify a UMTS bearer service. The UMTS BS manager will interrogate with its associated admission / capability control whether the network entity supports a specific requested service and whether the required resource is available. The UMTS BS manager at CN EDGE also has to verify with the subscription control the administrative rights for using the service. Based on the layered UMTS QoS architecture, UMTS bearer service manager will translate the UMTS bearer service attributes into attributes of the lower layer service manager. For example, the UMTS BS manager of the CN EDGE will translate the UMTS bearer service attributes into RAB service attributes, Iu bearer service attributes, and CN bearer service attributes. Each low layer will provide service to upper layer service manager.

2.3.2 QoS management in user plane

The Figure 4 shows the QoS management functions of UMTS bearer service in the user plane. They are mapping function, classification function, resource manager and traffic conditioner. They are used to maintain the data transfer characteristics according to the commitments established by the UMTS BS control functions.

Mapping function provides each data with the specific marking for receiving the requested QoS at the transfer. It is located at UTRAN, Gateway.

Classification function (Class.) in the MT and Gateway assigns user data units received from the external bearer service or the local bearer service to the appropriate UMTS bearer service according to the QoS requirement of each user data unit.

Traffic conditioner provides conformance between the negotiated QoS for a service nad the data unit traffic. Policing or traffic shaping is used for traffic conditioning. The policing function compares the data unit traffic with the related QoS attributes. Data units not matching the relevant attributes will be dropped or marked as not matching, for preferential

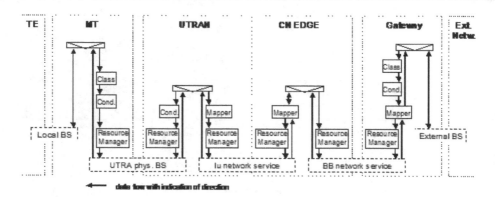

Fig. 4. QoS management function for UMTS bearer service in the user plane

dropping in case of congestion. The traffic shaper forms the data unit traffic according to the QoS of the service. The shaper algorithm is "Token Bucket Algorithm". At MT side, the traffic conditioner (Cond.) provides conformance of the uplink user data traffic with the QoS attributes of the relevant UMTS bearer service. In the Gateway a traffic conditioner may provide conformance of the downlink user data traffic with the QoS attributes of the relevant UMTS bearer service; i.e., on a per PDP context basis. A traffic conditioner in the UTRAN forms this downlink data unit traffic according to the relevant QoS attributes.

Resource Manager distributes the available resources between all services sharing the same resource. The resource manager distributes the resources according to the required QoS. Example means for resource management are scheduling, bandwidth management and power control for the radio bearer. It is located at MT, UTRAN, CN EDGE and Gateway.

2.4 QoS management in UMTS

2.4.1 QoS in CS domain

CS call control will control the QoS in the CS domain. MSC server and CS-MGW will provide QoS related functions. For UMTS release '99 CS-CC, the QoS related bearer definitions of GSM (as defined in bearer capability information element, octet 6 and its extensions) are sufficient.

In the CS domain the UE can only request a certain service with a well defined set of QoS parameters. CS domain uses traditional "circuit switching" technology, i.e. a constant set of resources exclusively dedicated to a connection. All CS domain services will require real-time bearers but differ in bandwidth and delay requirements. Based on the Bearer Capability information element the following services can be identified:

1. speech: from the Information Transfer Capability (ITC) parameter;

2. data, non-transparent: from the ITC and Connection element (CE) parameters;

3. data, transparent: from the ITC and CE parameters.

According to the standard, speech as well as the transparent data service is mapped to the conversational class while the non-transparent data service is mapped to the streaming class.

The MSC-Server is responsible for the service negotiation which includes subscription check and admission control. Furthermore, the QoS parameters corresponding to the service have to be mapped specifically for the interfaces to the UTRAN, GMSC-Server and CS-MGW. To provide QoS the CS-MGW has to perform admission control for the bearer resource which is therefore a part of the call admission control. Additionally, the CS-MGW is responsible for the QoS mappings to the Iu-, CN- and external bearer services.

With the separation of transport and control in the CS domain the resource allocation becomes more flexible. The new transport techniques ATM and IP (which are available for the CS bearer independent domain) allow a more efficient network usage from a parallel transmission of voice and data possibly leading to the consolidation of the whole PLMN (including the PS domain and parts of RAN) on one transport network. The QoS issue in the CS domain with IP or ATM based transport is to guarantee the same QoS as a TDM based PLMN with increased bandwidth efficiency.

2.4.2 QoS in PS domain

Since the PS domain provides packet data services, which are characterized by individual transmission of packets. QoS of different packet service is defined by a set of explicitly defined QoS parameters. So some effort is necessary to assure that packets of one flow are transmitted with guaranteed QoS.

The 3GPP specificationscitep (3GPP23107, 2011) define the QoS management functions in the UMTS bearer service for both control plane and user plane. Establishment of QoS within a UMTS network is achieved through the Packet Data Protocol (PDP) context activation procedure. The user equipment (UE) sends an Active PDP Context Request message to the SGSN, which contains the desired QoS profile, among other parameters. With these QoS attributes the treatment of the packets is sufficiently defined and all packets (or flows) belonging to the same PDP context are handled in the same way by the GPRS bearer service. After the UE sends a PDP context request with explicitly defined QoS parameters, the SGSN will negotiate the QoS parameters which includes subscription check and admission control (capability and resource check). Then the SGSN interacts with the UTRAN and the GGSN to establish the PDP context. The GGSN also performs admission control, i.e. the resource check for the GPRS as well as for the external bearer service. Additionally, the GGSN has to map QoS parameters from the GPRS to the external bearer service.

2.4.3 QoS in IP multimedia subsystem

IP multimedia subsystem (IMS) is introduced in 3GPP Release 5. It is an IP based system overlay on the PS domain. It support Session Initiation Protocol (SIP) based multimedia service. IMS can support end-to-end IP QoS service by using IP based bearer service. The IP based bearer service is supported by MS local bearer service, UMTS bearer service and external service.

Since 3GPP Release 5, the UMTS will support QoS in the IP layer between UE and multimedia application server/UE. The UE and GGSN have important roles in the IP layer QoS framework, they map QoS parameters between IP layer bearer service and UMTS bearer service. The detail will be discussed in the section 3.

For supporting IP layer QoS, 3GPP introduces the policy based QoS management in the IMS. The policy framework is recommended for policy management in IETF. The detail discusses is given in the section 4.

3. End-to-end IP QoS over UMTS

With the evolution of the 3GPP standards, operators want to provide end-to-end QoS enabled services in UMTS. The end-to-end behavior provided by a series of network elements is an assured level of bandwidth that produces a delay-bounded service with no queueing loss for all conforming packet data (RFC2212, 1997). Assuming the network is functioning correctly, these applications may assume that (?):

- A very high percentage of transmitted packets will be successfully delivered by the network to the receiving end-nodes. (The percentage of packets not successfully delivered must closely approximate the basic packet error rate of the transmission medium).

- The transit delay experienced by a very high percentage of the delivered packets will not greatly exceed the minimum transmit delay experienced by any successfully delivered packet. (This minimum transit delay includes speed-of-light delay plus the fixed processing time in routers and other communications devices along the path.)

The end-to-end QoS architecture is provided in Figure 1 in section 2. IP level mechanisms are necessary in providing end-to-end QoS services by interacting TE/MT local bearer service, GPRS bearer service and external bearer service. In this section, how to implement end-to-end IP QoS is described.

3.1 QoS mechanisms in IP

Quality of service refers to the nature of the packet delivery service provided, as described by parameters such as achieved bandwidth, packet delay, and packet loss rates (RFC2216, 1999). The Internet, as originally conceived, offers only a very simple quality of service (QoS), point-to-point best-effort data delivery. It means the network just offered available bandwidth and delay characteristics dependent on instantaneous network load. Before real-time applications such as remote video, multimedia conferencing, visualization, and virtual reality can be broadly used, the Internet infrastructure must be modified to support real-time QoS, which provides some control over end-to-end packet delays. From the view of applications, QoS is realized by adequate provisioning of the network infrastructure. In contrast, a network with dynamically controllable quality of service allows individual application sessions to request network packet delivery characteristics according to their perceived needs, and may provide different qualities of service to different applications. There are two basic types of QoS available (qodwhitepaper, 1999):

- Resource reservation (integrated services): network resources are apportioned according to an application's QoS requirement, subject to bandwidth management policy.

- Prioritization (differentiated services): network traffic is classified and apportioned network resources according to bandwidth management policy.

The both types of QoS can be applied to individual application 'flow' or to flow aggregates, so there are two other methods to characterize types of QoS:

- Per flow: A 'flow' is defined as an individual, uni-directional data stream between two clients (caller and callee), uniquely identified by a 5-tuple (transport protocol, source address, source port number, destination address, and destination port number).

- Per aggregate: An aggregate is simply two or more flows. Usually the flows have something in common (e.g. any one or more of 5-tuple parameters, a label or a priority number, or perhaps some authentication information).

Generally, we can see that there are two methods to support QoS in IP network. One is IntServ (Integrated Service), the other is DiffServ (Differentiated Service). IneServ is Per Flow based QoS control mechanism. Diffserv is Per Aggregate based QoS control mechanism. To accommodate the need for these two types of QoS, there are following QoS protocols and algorithms:

- ReSerVation Protocol (RSVP):

- Differentiated Service (DiffServ)

- Multi Protocol Labeling Switching (MPLS)

3.1.1 IntServ

The Internet integrated services (IntServ) framework provides the ability for applications to choose among multiple, controlled levels of delivery service for their data packets. It can provide hard QoS guarantee to individual traffic flows. To support this capability, two things are required (?):

- Individual network elements (subnets and IP routers) along the path followed by an application's data packets must support mechanisms to control the quality of service delivered to those packets.

- A way to communicate the application's requirements to network elements along the path and to convey QoS management information between network elements and the application must be provided.

In the integrated services framework the first function is provided by QoS control services such as Controlled-Load (RFC2211, 1997) and Guaranteed (RFC2212, 1997). The second function may be provided in a number of ways, but is frequently implemented by a resource reservation setup protocol such as RSVP (RFC2205, 1997).

The controlled load service is intended to support a broad class of applications which have been developed for use in today's Internet, but are highly sensitive to overloaded conditions. Important members of this class are the "adaptive real-time applications" currently offered by a number of vendors and researchers. These applications have been shown to work well on unloaded nets, but to degrade quickly under overloaded conditions. It is equivalent to "best effort service under unloaded conditions". The controlled-load service is intentionally minimal, in that there are no optional functions or capabilities in the specification. The service offers only a single function. It is better than best effort, but cannot provide strictly bounded service as guaranteed service.

The controlled-load service can be implemented by using evolving scheduling and admission control algorithms. The implementations are highly efficient in the use of network resources.

Guaranteed service guarantees that datagrams will arrive within the guaranteed delivery time and will not be discarded due to queue overflows, provided the flow's traffic stays within its specified traffic parameters. It is similar to emulate a dedicated virtual circuit. This service is intended for applications which need a firm guarantee that a datagram will arrive no later than a certain time after it was transmitted by its source. For example, some audio and video "play-back" applications are intolerant of any datagram arriving after their play-back time. Applications that have hard real-time requirements will also require guaranteed service.

Guaranteed service does not attempt to minimize the jitter (the difference between the minimal and maximal datagram delays); it merely controls the maximal queueing delay. Because the guaranteed delay bound is a firm one, the delay has to be set large enough to cover extremely rare cases of long queueing delays. Several studies have shown that the actual delay for the vast majority of datagrams can be far lower than the guaranteed delay. Therefore, authors of playback applications should note that datagrams will often arrive far earlier than the delivery deadline and will have to be buffered at the receiving system until it is time for the application to process them.

Guaraneteed service represents one extreme end of delay control for networks. Most other services providing delay control provide much weaker assurances about the resulting delays. In order to provide this high level of assurance, guaranteed service is typically only useful if provided by every network element along the path (i.e. by both routers and the links that interconnect the routers). Moreover, as described in the Exported Information section, effective provision and use of the service requires that the set-up protocol or other mechanism used to request service provides service characterizations to intermediate routers and to the endpoints.

Integrated Services routers uses admission control and resource allocation method to offer QoS guarantee. A token-bucket model is used to characterize the input/output queueing algorithm. It can smooth the flow of outgoing traffic. The IntServ parameters include (qodwhitepaper, 1999):

Token rate (r): The continually sustainable bandwidth (bytes/second) requirement for a flow. It represents the average data rate into the bucket, and the target shaped data rate out of the bucket.

Token-bucket rate (b) : the extent to which the data rate can exceed the sustainable average for short periods of time, or the amount of data sent cannot exceed rT+b (where T is any time period).

Peak rate (p): It is the maximum send rate (bytes/second) if known and controlled. At any time period (T), the amount sent data cannot exceed M+pT.

Minimum policed size (m): The size (byte) of the smallest packet (data payload only) can be generated by the sending application. The size m is not an absolute number. If the percentage of small packets is small, the number m should increased to reduce the overhead estimate. All packets smaller than m are treated as size m.

Maximum packet size (M): The biggest size of a packet (bytes). The M is absolute number. Any packets (size > M) are considered out of spec and may not receive QoS controlled service.

3.1.2 RSVP

For offering IntServ, a way to communicate the application's requirements to network elements along the path and to convey QoS management information between network elements and the application must be provided. A resource reservation setup protocol called RSVP (rfc2205, 1997) is implemented for this purpose. It is a signaling protocol that can provide reservation setup and control to enable the integrated services by using a variety of QoS control, a variety of setup mechanisms.

A simplified RSVP working flow is shown in Figure 5

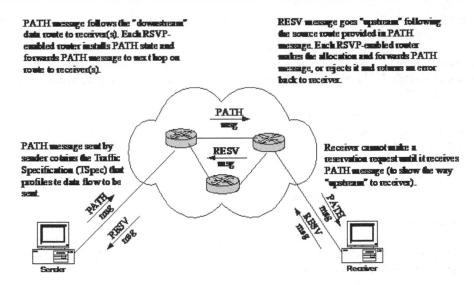

Fig. 5. RSVP setup flow

1. When a sender wants to set up a traffic link, it will generate the traffic specification (TSpec), which describes data traffic, such as upper/lower bounds of bandwidth, delay, and jitter. Then RSVP sends out a PATH message containing TSpec to the receiver(s) (unicast or multicast). Along the route, each RSVP-enabled router trigger a "path-state" that includes the previous source address of the PATH message.

2. After receiver receives the PATH message, the receiver sends a RESV message "upstream" to make a resource reservation. The RESV message includes a request specification (RSpec) which indicates what type of IntServ required âÅŞ either Controlled Load or Guaranteed, a filter specification (filter spec) (indicating e.g. the transport protocol and port number). The RSpec and filter spec represent a flow-descriptor that RSVP routers use to identify each reservation.

3. Along the RSVP upstream, RSVP routers use the admission control to authenticate the resource reservation request and allocate the necessary resources when the routers receive the RESV message. If the request cannot be met (due to no enough bandwidth or authorization failure), the RSVP router returns an error back to the receiver. If the request is accepted, the router forwards the RSVP message to the next router.

4. When the last router receives the RESV message and accepts the request, it sends out a confirmation message back to the receiver.

5. There is an explicit tear-down process for a reservation when sender or receiver terminate a RSVP session.

3.1.3 DiffServ

The Integrated Services/RSVP model relies upon traditional datagram forwarding in the default case, but allows sources and receivers to exchange signaling messages which establish additional packet classification and forwarding state on each node along the path between them (rfc1633, 1994). In the absence of state aggregation, the amount of state on each node scales in proportion to the number of concurrent reservations, which can be potentially large on high-speed links. This model also requires application support for the RSVP signaling protocol. Differentiated service is a simple method by classifying services of different applications (rfc2475, 1998). Currently there are two standard per hop behaviour (PHBs) define two traffic classes:

- Expedited Forwarding (EF): Has a single codepoint (DiffServ value). Ef minimize delay and jitter and provides the highest level of aggregate quality of service. Any traffic that exceeds the traffic profile is discarded (?). EF class offers a low jitter, low delay service. UserâĂŹs traffic cannot exceed the agreed peak rate. Otherwise, the packets will be discarded.

- Assured Forwarding (AF): Has four classes and three drop-precedence within each class (a total of twelve codepoints). Excess AF traffic is not delivered with as high probability as the traffic "within profile", which means it may be demoted but not necessarily dropped (?). The AF class is suitable for delay-tolerant applications. The guarantee just implies that the better QoS class will give a better performance than the low-level QoS class. Network operator can define their own per-hop behavior.

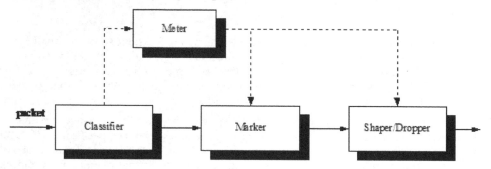

Fig. 6. DffServ architecture

DiffServ offers a simple QoS management method without signaling mechanism. The DiffServ architecture is shown in Figure 6. It includes classifier and traffic conditioner. A traffic conditioner contains the following elements: meter, maker, shaper/dropper. The differentiated services architecture is based on a simple model where traffic entering a network is classified and possibly conditioned at the boundaries of the network, and assigned to different behavior aggregates. Each behavior aggregate is identified by a single DiffServ codepoint (DSCP). Within the core of the network, packets are forwarded according to the per-hop behavior associated with the DiffServ codepoint.

When a traffic flow enters a DiffServ network, the flow is selected by a classifier, which steers the packets to a logical instance of a traffic conditioner. A meter is used to measure the traffic flow agains a traffic profile. A meter is used (where appropriate) to measure the traffic stream against a traffic profile. The state of the meter with respect to a particular packet (e.g., whether it is in- or out-of-profile) may be used to affect a marking, dropping, or shaping action. When packets exit the traffic conditioner of a DS boundary node the DiffServ codepoint of each packet must be set to an appropriate value.

3.1.4 MPLS

MPLS is a key development in IETF that will add a number of essential capabilities to today's best effort IP networks, including

- Traffic Engineer, enhancing overall network utilization by creating a uniform or differentiated distribution of traffic throughout the network.
- Providing traffic with different Classes of Service (CoS)
- Providing traffic with different Quality of Service (QoS)
- Supporting network scalability, providing IP based Virtual Private Networks (VPN)

MPLS borrows the idea from ATM switching. It remains independent of the Layer-2 and Layer-3 protocols. Besides IP, other network protocols (such as IPX, ATM, PPP or Frame-Relay) also can work with MPLS. MPLS resides on routers. When a packet flow enters a edge router of the MPLS domain, all packets are marked to clarify priority with a fixed-length label (20 bits label). The label identifies the packets routing information in this MPLS network, also define the quality of service for the packets.

A MPLS domain includes label edge routers (LERs) and label switching routers (LSRs). The route taken by an MPLS-labeled packet is called the label switched path (LSP). LST is a high-speed router in the core of a MPLS network, which participates in the establishment of LSPs. LER is a router that operates at the edge of a MPLS network. It is used to assign and remove labels when packets enter or exit the MPLS network.

MPSL is similar to DiffServ because it also marks traffic at ingress of a MPLS network, and un-marks at egress gate. However, MPLS marking is used to decide the next hop router while DiffServ marking is used to determine priority in route itself.

3.2 QoS management functions for end-to-end IP QoS

This section describes how to provide Quality of Service in UMTS for the end-to-end services through the TE/MT local bear service, GPRS bearer service and external bearer service shown in the Fig. 1. To provide end-to-end IP QoS, it is necessary to manage the QoS within each domian. An IP BS Manager is used to control the external IP bearer service. Due to the different techiniques used within the IP network,this communicates to the UMTS BS manager through the Translation function.

At PDP context setup the user shall have access to one of the following alternatives, basic GPRS IP connectivity service or enhanced GPRS based services. To enable coordination between events in the application layer and resource management in the IP bearer layer, a logical element, the Policy and Charging Rules Function (PCRF), is used as a logical policy decision element which will be detailed in section 4. It is also possible to implement a policy decision element internal to the IP BS Manager in the GGSN. While interworking with the external network, the RSVP, DiffServ, MPLS will be used.

QoS management functions is shown in Fig. 7 which describes how to control the external IP bearer services and how they relate to the UMTS bearer service QoS management entity.

IP BS Manager uses standard IP mechanisms to manage the IP bearer services. These mechanisms may be different from mechanisms used within the UMTS, and may have different parameters controlling the service. When implemented, the IP BS Manager

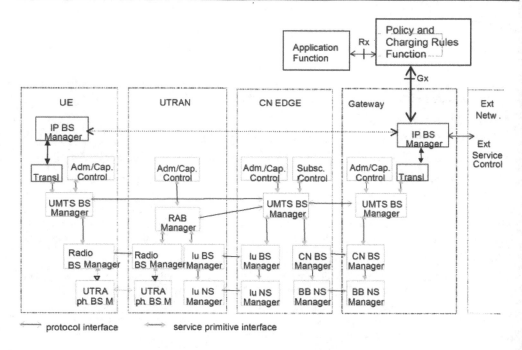

Fig. 7. QoS management functions for end-to-end QoS in UMTS

may include the support of DiffServ Edge Function and the RSVP function. The Translation/mapping function provides the inter-working between the mechanisms and parameters used within the UMTS bearer service and those used within the IP bearer service, and interacts with the IP BS Manager. In the GGSN, the IP QoS parameters are mapped into UMTS QoS parameters, where needed. In the UE, the QoS requirements determined from the application layer (e.g., SDP) are mapped to either the PDP context parameters or IP layer parameters (e.g., RSVP). If an IP BS Manager exists both in the UE and the Gateway node, it is possible that these IP BS Managers communicate directly with each other by using relevant signalling protocols. The required options in the table define the minimum functionality that shall be supported by the equipment in order to allow multiple network operators to provide interworking between their networks for end-to-end QoS. Use of the optional functions listed below, other mechanisms which are not listed (e.g. over-provisioning), or combinations of these mechanisms are not precluded from use between operators. The IP BS Managers in the UE and GGSN provide the set of capabilities for the IP bearer level as shown in table 2. Provision of the IP BS Manager is optional in the UE, and required in the GGSN.

Capability	UE	GGSN
DiffServ Edge Function	Optional	Required
RSVP/IntServ	Optional	Optional
IP Policy Enforcement Point	Optional	Required

Table 2. IP BS Manager capability in the UE and GGSN

4. Policy based QoS management - IP QoS for IMS

This section will provide an overview of policy based QoS management in UMTS IMS. Although the UMTS packet switched (PS) domain can support IP QoS enabled multimedia applications, there are many ways of establishing QoS guaranteed IP multimedia session through a signaling protocol before it can map and reserve the equivalent amount of QoS resources along the data path in the PS domain. In order to support interoperation among UMTS network providers, the IP multimedia Subsystem (IMS) is standardized by the 3GPP to serve Session Initiation Protocol (SIP) signaled IP multimedia services over the UMTS PS domain.

The central problem of providing consistent end-to-end IP QoS services is the difficulty of configuring the network devices like routers and switches to handle packet flows in a manner that satisfies the requested QoS requirements. Policy-based QoS management is used to control QoS resources in the UMTS IMS.

4.1 Introduction to policy-based QoS network

4.1.1 The need for policy based network

There is a consistent effort to implement new IP multimedia services in UMTS. While the IP based network is well suited for packet data transfer, providing consistent end-to-end IP QoS services is the difficulty of configuring the network devices like routers and switches to handle packet flows in a manner that satisfies the requested QoS requirements. This problem is especially acute when the end-to-end data path of an IP QoS session crosses multiple administrative domains managed by different operators. Although the operators agree on the QoS requirements of a particular set of IP services, they may not configure their network devices in the same way to implement the services due to differences in the network topologies, QoS mechanisms available in the network devices and non-technical management requirements. Thus, there is a need to create a solution that permits network operators, including UMTS network operators, to easily configure their networks to implement consistent IP QoS services without dealing with the complexity of their networks.

Policy-based Networking (PBN) is a novel approach to configure myriad network devices in an administrative domain to implement a set of IP QoS services. Policy-based network will allow the network operator to define, in a succinct and organized fashion, operator policies that automatically effect change on specific equipment in the network environment. The end result is that the end-to-end network performance will meet the general expectations of UMTS service provider environment.

4.1.2 What is policy?

A policy is a set of business rules that guide and determine how to manage network resources. The basic concept is that policy rule(s) describe how network to act when specific condition(s) happen. "Policy" can be defined from two perspectives: (POLICYTERM, 2001). - A definite goal, course or method of action to guide and determine present and future decisions. "Policies" are implemented or executed within a particular context (such as policies defined within a business unit). - Policies as a set of rules to administer, manage, and control access to network resources. [RFC3060] Note that these two views are not contradictory since individual rules may be defined in support of business goals.

Policy can be represented at different levels, ranging from business goals to device-specific configuration parameters. Enforcement of policy ensures that business rules are always followed. Policy rule is a basic building block of a policy-based system. It is the binding of a set of actions to a set of conditions - where the conditions are evaluated to determine whether the actions are performed. [RFC3060] A condition is a set of expressions or objects used to determine whether a given policy rule's action should be performed. A condition answers the question, "when and where do we enforce a policy?" An action defines what to be done to enforce a policy rule, when the conditions of the rule are met. Policy actions may result in the execution of one or more operations to affect and/or configure network traffic and network resources. An action answers the question, "what must be done to enforce a policy?"

A policy also defines how the network's resources are to be allocated among its clients. Clients can be individual users, departments, host computers, or applications. Resources can be allocated based on time of day, client authorization priorities, availability of resources, and other factors. How resources are allocated can be static or dynamic (based on variations in traffic). Policies are created by network managers and stored in a repository. During network operation, the policies are retrieved and used by network management software to make decisions.

4.1.3 Policy framework & architecture

The network operators negotiate Service Level Agreements (SLAs) that describe the sets of IP QoS services that they have mutually contracted to provide. Individual operators will then transform the QoS requirements specified in the SLAs into sets of policy rules that will be applied to their network domains to implement the contracted IP QoS services. The IETF has defined a policy framework (RFC2753, 2001) as shown in Figure 8 to transform the sets of policy rules to network device configurations in an administrative domain. The sets of policy rules are stored in the Policy Repository through the Policy Management Tool. The Policy Decision Point (PDP) retrieves the appropriate policy rules from the Policy Repository in response to policy events that are triggered by the contracted IP QoS services, e.g., the reception of an RSVP message by the Policy Enforcement Point (PEP). It translates the acquired policy rules into a set of QoS mechanism configuration actions that is communicated to the PEP as policy decisions. The PEP then executes the actions spelt out in the supplied decisions to handle the triggering policy events in accordance with the requested IP QoS services. Alternatively, the retrieved policy rules may be returned to the PEP, which is capable of translating them into configuration actions. These policy rules can be cached in the PEP so that similar future triggering policy events can be serviced locally without further interactions with the PDP.

Outsourcing and Provision Model in PBN

There are two main models for policy management: outsourcing and provisioning. The outsourcing model assumes there is a signaled event in the Policy Enforcement Point (PEP) that must be resolved based on policy criteria. The PEP outsources the decision-making to an external policy decision point (PDP). This outsourcing model is sometimes referred to as "Pull" mode, or "reactive" mode, since the PEP pulls policy decisions from the PDP, while the PDP responds according to the PEP events.

The provisioning model is almost the mirror image of the outsourcing model. In this system, the PDP predicts future configuration needs, and proactively provisions resources accordingly. In other words, rather than responding to PEP events, the PDP prepares

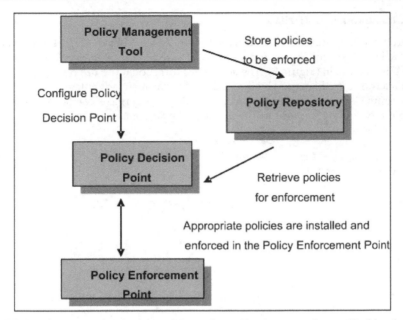

Fig. 8. A PBN architecture that is derived from the policy framework specified by the IETF

and "pushes" configuration information to the PEP. This takes place as a result of external events (unrelated to the PEP) such as change of applicable policy, time of day, expiration of account quota, or information from third party (non-PEP) signaling.

Both models employ policy servers as the PDP to control the network devices that enforce the policy (i.e. PEPs). PBN also offers a policy repository for storing policy information accessed by the PDPs in the system. To communicate policy information between PDPs and PEPs, the COPS policy protocol is engaged. Additionally, the LDAP protocol functions to access the policy repository.

Policy Decision Point (PDP)

The PDP is the PBN component that directly controls the network devices or policy enforcement points (see next section). Functionally, the PDP handles policy information that has been entered into the PBN management system. The policy data used by the PDP can either be obtained in real-time upon entry into the management console, or from the policy repository on an as-needed basis. The function of the PDP involves retrieving policy, interpreting policy, detect policy conflicts, receiving policy decision requests from PEPs, determining which policy is relevant, applying the policy and returning the results. It also sends policy elements the PEP.

Policy Enforcement Point (PEP)

Network devices that receive and enforce the decisions from the PDP are referred to as PEPs. In both outsourcing and provisioning policy management models, PEPs receive policy decisions and enforce them at the packet level as data passes through the devices.

4.2 Policy framework in UMTS IMS

To support IP based multimedia services, the IP Multimedia Subsystem (IMS) is introduced in the 3GPP Release 5 specifications. It provisions IP based multimedia services as an extension of the UMTS PS domain (Figure 9). The added IMS functionalities are control functionalities; the user data traffic is still carried by the PS domain. The main advantage of the IMS is that it offers operators a scalable service platform on which new services can be developed rapidly in a flexible way, without requiring any change to the PS domain.

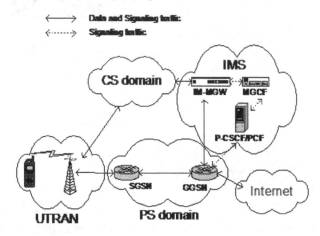

Fig. 9. A simple UMTS network with IMS

Having put in place the functionalities to handle IP multimedia calls, the next big challenge is to ensure that sufficient QoS resources are provided to authorized users in the UMTS network. A policy-based QoS solution is adopted by the 3GPP for this purpose.

As mentioned in section 4.1.3, the reference model of a policy-based network consists of two main elements, the PDP and the PEP (RFC2753, 2001). PEPs often reside in policy aware network nodes that carry out actions stipulated by policy rules. The actions taken are based on the decisions of a PDP, which retrieves the policy rules from a repository. The PDP is the final authority, which the PEP needs to refer to for actions to be taken.

In the IMS, the Policy and Charging Rules Function (PCRF) (3G23203, 2008) plays the role of the PDP and online charging and offline charging functions, the Policy and Charging Enforcement Functions plays the role of the PEP. Policy charging and rules function (PCRF) is the node designated in real-time to determine policy rules in a multimedia network. As a policy tool, the PCRF plays a central role in WCDMA networks. Unlike earlier policy engines that were added on to an existing network to enforce policy, the PCRF is a software component that operates at the network core and efficiently accesses subscriber databases and other specialized functions, such as a charging systems, in a scalable, reliable, and centralized manner. The PCRF as the part of the network architecture that aggregates information to and from the network, operational support systems, and other sources (such as portals) in real time, supporting the creation of rules and then automatically making intelligent policy decisions for each subscriber active on the network. Such a network might offer multiple services, quality of service (QoS) levels, and charging rules. In this chapter, we will focus on policy based management functions.

The PCRF communicates with the PCEF via the Gx interface (3G29212, 2008). It allows two modes of operation. In the "push" mode, the PCRF initiates communication with the PCEF and sends the PCEF its decision. In the "pull" mode, the PCEF initiates communication with the PCRF to request a decision for a particular IP flow. The Gx interface and the protocol used for communication on the interface are described in the following.

Figure 10 depicts the relationship between these entities. In the following subsections, each of these network elements will be described.

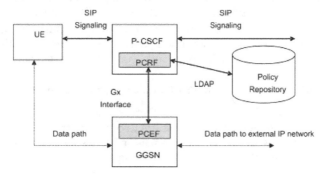

Fig. 10. Policy architecture in UMTS IMS

PCRF in Proxy-CSCF

During the establishment of a SIP session, a P-CSCF is the first contact point in the IMS domain for a UE [3G23228]. Hence it is the natural place to authorize the usage of network resources such as the bandwidth requested by the UE. The QoS requirements of the UE are carried in the Session Description Protocol (SDP) description within a Session Initiation Protocol (SIP) message. Besides the QoS requirements in the SDP description, the PCRF also examines the source and destination IP addresses and port numbers in its decision-making. The PCRF refers to the policy rules, which are generally stored in a policy repository, governing the local domain. It then generates an authorization token that uniquely identifies the SIP session across multiple Packet Data Protocol (PDP) contexts terminated by a GGSN. This token is sent to the UE via SIP messages so that the UE can use it to identify the associated session flows to the PCEF in the GGSN in subsequent transmission of IP packets. This mechanism is consistent with the IETF specification on supporting media authorization in the SIP protocol [RFC3313]. The flow of events in session set-up is described in Section 4.6.

PCEF in Gateway GPRS Support Node (GGSN)

In the PS domain, a GGSN maintains connectivity to other packet switched external networks such as the Internet. From the service point of view, the GGSN controls which IP flows are permitted into the external IP network by policing the IP packets based on their source and destination IP addresses and port numbers [3G23228]. As such, it is logical to embed the PCEF in the GGSN. The role of the PCEF is to ensure that only authorized IP flows are allowed to use network resources that have been reserved and allocated to them.

The policy enforcement function in the GGSN is called a "gate". A gate comprises a packet classifier, a traffic meter, and the relevant packet handling mechanisms for packets that have been matched by the packet classifier. When an IP flow is authorized by the PCRF to use the specified network resources, the PCEF opens the "gate" for the flow and effectively commits the network resources to the flow by allowing it to pass through the packet handling mechanisms (i.e., policing or marking). On the other hand, if an IP flow is not permitted by the PCRF to use the requested resources, the PCEF closes the âĂIJgateâĂİ and drops the IP packets of the flow. This process is called policy-based admission control. It ensures that an IP flow is only allowed to use resources that have been approved by the policy rules. The above process takes place at the IP bearer service (BS) level. The translation/mapping function within the GGSN will map this resource information into the format used by the admission control function at the UMTS BS level.

The PCEF may store decisions in a local policy decision point, thus allowing the GGSN to make the admission control decisions without additional interactions with the PCRF. This will reduce the traffic over the Gx interface and lessen the processing load on the PCRF.

4.3 Policy-based QoS delivery: an example of policy based call control

There are several reasons why a policy-based QoS framework is adopted for the UMTS. Policy-based QoS control allows network operators to configure their network devices easily. It provides a high level view of the network devices and allows the automated translation of business level policies to suitable information for configuring network devices.

UMTS requires a strict authorization of users so that the network resources are not abused. Once authorized and approved, the UMTS must guarantee that the resources are made available to the legitimate users. If these requirements are not met, these users may be denied the use of the resources, leading to dissatisfaction with the quality of service provided. To ensure that this is not the case, all IP multimedia calls must go through the following steps:

1. Authorization of resources;
2. Reservation of resources. This is to make sure that the resources are available when the "phone" rings;
3. Once the called party picks up the "phone", the network resources reserved previously are committed. The charging process is then triggered.

In all these steps, policy rules are used in approving the requests, and the PCRF is the sole approving authority. By changing the policy rules in the PCRF, a network operator can alter the IP multimedia services it offers to its subscribers without having to know the details of its network configuration and the types and mechanisms of the network devices.

To meet the above requirements, two procedures are needed for the establishment of an IP multimedia session in addition to the normal GPRS bearer establishment procedures. These procedures are Authorize QoS Resources and Approval of QoS Commit (3G29212, 2008). Similarly, the procedures, Removal of QoS Commit and Revoke Authorization of QoS Resources, are carried out to reverse the authorization and commitment of QoS resources when an IP multimedia session is terminated. The following provides an overview of the session set-up procedures, in particular, the emphasis is on the additional procedures introduced by the service-based local policy.

4.4 Session establishment procedures

The establishment of an IP multimedia service session with policy control differs from that without policy control in that additional steps are taken to check the policy rules for a decision on whether to grant or deny the required network resources to the session. As the signaling messages used to set up the session take a different path from that used for the data flow, an authorization token and a flow identifier are used to associate the session with its IP data flow (UMTSGO01, 2001). The GGSN, which is located on the data path, relies on this binding information to enforce the policy rules on the IP data flows.

Figure 11 depicts the sequence of events that take place during the establishment of an IP multimedia service session. Note that a number of signaling messages have been omitted for clarity. The events are described in the following paragraphs:

Steps 1-5: The UE sends a session set-up request (i.e., SIP INVITE) to the P-CSCF indicating, among other things, the media streams to be used in the session. This message is routed to the called party via a number of other CSCFs (viz., the caller and callee S-CSCFs) along the signaling path. The S-CSCFs perform the appropriate session control services for the UEs. In particular, they maintain a session state that is needed by the network operator to support the requested service.

Steps 6-14: The called party responds with a provisional SIP 183 response message. This message is routed to the calling party via the same CSCFs along the (reversed) signaling path. When the callee P-CSCF receives this message, it examines the SDP description within the message to determine the QoS parameters requested for the session. The P-CSCF sends the necessary information in this SIP message (e.g., the bandwidth, IP addresses and ports, etc.) to the PCRF for authorization of the session request. If the policy permits, the PCRF responds with an authorization token that can be used to identify the authorized session and resources. The P-CSCF includes the token in the response (SIP 183 message) and forwards it to the callerâĂŹs UE. A similar process is carried out at the caller P-CSCF when it receives the SIP 183 message. This process of authorization by the PCRF and the generation of a token is called "Authorize QoS Request".

Steps 15-22: In between steps 14 and 15, other message exchanges take place between the caller and the callee. However, these are not important in this particular example and are omitted for clarity. The caller's UE starts the resource reservation by sending a PDP Context Activation Request to the GGSN. The authorization token and the flow identifier(s) from the PCRF are included to identify the IP data flow(s) of the session. When the GGSN receives the PDP Context Activation Request, it sends a policy decision request to the PCRF to determine whether the resource reservation request should be accepted. The PCRF uses the token in the message to correlate the request for resources with the media authorization previously granted to the session. The PCRF then sends a decision to the GGSN. If the PCRF approves the resource reservation, the GGSN sends a PDP Context Activation Response to the UE indicating that the resource reservation has been completed. A similar process takes place at the callee's end.

Steps 23-31: In between steps 22 and 23, there are other events, e.g., 180 Ringing, that take place. These events are omitted to prevent cluttering Figure 3-22. When the callee answers the call, a SIP 200 OK message is sent towards the caller. When the SIP 200 OK reaches the P-CSCF, it will approve the QoS commitment by sending a decision to the GGSN. Upon receiving this message, the GGSN opens the gate, thereby effectively permitting the IP data flow to use the resources reserved previously. Once this is done, the GGSN responds to the PCRF with a report on the status of the session. A similar process takes place at the caller's end. When this entire process is completed, the proper resources on the data path have been reserved and committed to the session.

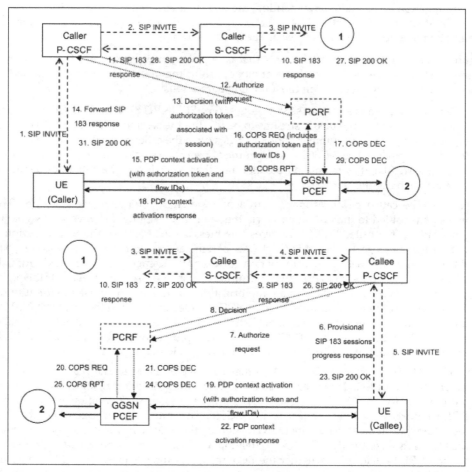

Fig. 11. Session authorization mechanism in a UE-to-UE session establishment process

5. References

Nortel White Paper (2002). Benefits of Quality of Service (QoS) in 3G Wireless Internet, Nortel Networks.

Sudhir Dixit, Yile Guo, Zoe Antoniou (2001). Resource management and quality of service in third-generation wireless network, *IEEE Communication Magazine*, Feb. 2001, pp.125-133.

Sotiris I. Maniatis, Eugenia G. Nikolouzou, & Iakovos S. Venieris (2002). QoS issues in the converged 3G wireless and wired networks, *IEEE Communications Magazine*, Aug. 2002, pp.44-53.

3GPP TS 23.107 (V9.2.0) (2011). Quality of Service(QoS) Concept and architecture (Release 9).

S.Shenker, C.Partridge, R.Guerin (1997), Specification of Guaranteed Quality of Service, *RFC2212*, Sept.1997.

J. Wroclawski (1997), Specification of the Controlled-Load Network Element Service, *RFC2211*, Sept. 1997.

S.Shenker, J.Wroclawski (1999), Network Element Service Specification Template, *RFC2216*, Sept. 1999.

White paper (1999), QoS Protocol & Architecture, *www.qosforum.com*, July, 1999.

R. Braden, D. Clark, S. Shenker, Integrated Services in the Internet Architecture: an Overview, *RFC 1633*, June 1994.

Braden, B., Ed., et. al., Resource Reservation Protocol (RSVP) - Version 1 Functional Specification, *RFC 2205*, September 1997.

S. Blake, D. Black, M. Carlson, E. Davies, Z. Wang, W. Weiss, An architecture for differentiated services, *RFC 2475*, Dec. 1998.

V. Jacobson, K. Nichols, K. Poduri, An expedited forwarding PHB, *RFC 2598*, June, 1999.

J. Heinanen, F. Baker, Weiss, J. Wroclawski, Assured forwarding PHB group, *RFC 2597*, June 1999.

White paper, Introduction to QoS Policies, *www.qosforum.com*, 1999

Survey on Policy-based networking, *INTAP*.

A. Westerinen, etc. Terminology for Policy-Based Management, *<draft-ietf-policy-terminology-04.txt>*, July, 2001.

B.Moore, E. Ellesson, J. Strassner and A. Westerinen, Policy Core Information Model – Version 1 Specification, *RFC 3060*, IETF, Feb. 2001.

M. Handley, et al., SIP: Session Initiation Protocol, *Internet draft (work in progress)*, *<draft-ietf-sip-rfc2543bis-09.txt>*, Feb. 2002

3GPP TS 29.212 (version 8.3.0), Policy and Charging Control Over Gx Reference Point (Rel 8), Dec. 2008.

3GPP TS 23.228 (version 8.3.0), IP Multimedia Subsystem- Stage 2 (Rel 8), June 2008.

3GPP TS 23.203 (version8.3.0),Policy and Charging Control Architecture (Rel. 8), 2008.

W. Marshall, et al., Private SIP Extensions for Media Authorization, *RFC 3313*, Nov. 2001

D. Durham, J. Boyle, R. Cohen, S. Herzog, R. Rajan and A. Sastry, The COPS (Common Open Policy Service) Protocol, *RFC 2748*, IETF, Jan. 2000.

R.Yavatkar, D.Pendarakis, R.Guerin, A Framework for Policy-based Admission Control, *RFC 2753*, IETF, Jan. 2000.

R. Atkinson, Security Architecture for the Internet Protocol, *RFC 2401*, IETF, Aug. 1995

T. Dierks and C. Allen, The TLS Protocol Version 1.0, *RFC 2246*, IETF, Jan. 1999

K. Chan, D. Durham, S. Gai, S. Herzog, K. McCloghrie, F. Reichmeyer, J. Seligson, A. Smith and R. Yavatkar, COPS Usage for Policy Provisioning, *RFC 3084*, Mar. 2001

L-N. Hamer, K. Chan, H. Syed, H. Shieh and R. Zwart, COPS-PR for outsourcing in UMTS: UMTS Go PIB, *draft-hamer-rap-cops-umts-go-00*, IETF, Nov. 2001

B. Moore, L. Rafalow, Y. Ramberg, Y. Snir, J. Strassner, A. Westerinen, R. Chadha, M. Brunner and R. Cohen, Policy Core Information Model Extensions, *draft-ietf-policy-pcim-ext-06*, Nov. 2001

J. Jason, L. Rafalow and E. Vyncke, IPsec Configuration Policy Model, *draft-ietf-ipsp-config-policy-model-03*, IETF, July 2001

Y. Snir, Y. Ramberg, J. Strassner, R. Cohen and B. Moore, Policy QoS Information Model, *draft-ietf-policy-qos-info-model-04*, IETF, Nov. 2001

S. Blake, D. Black, M. Carlson, E. Davies, Z. Wang and W. Weiss, An Architecture for Differentiated Service, *RFC 2475, IETF*, Dec. 1998

R. Braden, D. Clark and S. Shenker, Integrated Services in the Internet Architecture: an Overview, *RFC 1633, IETF*, June 1994

R. Braden, Ed., L. Zhang, S. Berson, S. Herzog and S. Jamin, Resource ReSerVation Protocol (RSVP) – Version 1 Functional Specification, *RFC 2205, IETF*, Sept. 1997

B Moore, D. Durham, J. Strassner, A. Westerinen, W. Weiss and J. Halpern, Information Model for Describing Network Device QoS Datapath Mechanisms, *draft-ietf-policy-qos-device-info-model-06, IETF*, Nov. 2001

M. Wahl, T. Howes and S. Kille, Lightweight Directory Access Protocol (v3), *RFC 2251, IETF*, Dec. 1997

G. Good, The LDAP Data Interchange Format (LDIF) - Technical Specification, *RFC 2849, IETF*, June 2000

Ebata, M. Takihiro, S. Miyake, et al., Interdomain QoS Provisioning and Accounting, *INET 2000*, Yokohama, Japan, July 2000

K. Nichols, V. Jacobson, L. Zhang, A Two-bit Differentiated Services Architecture for the Internet, *RFC 2638*, July 1999.

B. Teitelbaum, P. Chimento, "QBone Bandwidth Broker Architecture", work in progress, http://qbone.internet2.edu/bb/bboutline2.html.

6

Web-Based Laboratory Using Multitier Architecture

C. Guerra Torres and J. de León Morales
Facultad de Ingenieria Mecánica y Eléctrica
Universidad Autónoma de Nuevo León
México

1. Introduction

Actuality, Internet provides a convenient way to develop a new communication technology for several applications, for example remote laboratories. The remote access to complex and expensive laboratories offers a cost-effective and flexible means for distance learning, research and remote experimentation. In the literature, some works propose platforms based on the Internet in order to access experimental laboratories; nevertheless it is necessary that the platform provides a good architecture, clear methodology of operation, and it must facilitate the integration between hardware (HW) and software (SW) elements. In this work, we present a platform based on "multitier programming architecture" which allows the easy integration of HW and SW elements and offers several schemes of tele-presence: teleoperation, telecontrol and teleprogramming.

The remote access to complex and expensive laboratory equipment represents an appealing issue and great interest for research, learning education and industrial applications. The range potentially involved is very large, including among others, applications in all fields of engineering (Restivo et al., 2009; Wu et al., 2008).

It is well known that several experimental platforms are distributed in different laboratories in the world, and all of them are on-line accessible through the Internet. Since those laboratories require specific resources to enable a remote access, several solutions for harmonizing the necessary software and hardware have been proposed and described. Furthermore, due to their versatility, these platforms provide user services which allow the transmission of information in a simply way, besides being available to many people, having many multimedia resources.

The potentiality of remote laboratories (Gomez & Garcia, 2007) and the use of the Internet, as a channel of communication to reach the students at their homes, were soon recognized (Basigalup et al., 2006; Davoli et al., 2006; Callangan et al., 2005; Imbre & Spong, 2006; Rapuano & Soino, 2005).

Several works based on remote experimentation, which are used as excellent alternatives to access remote equipment, have been published (Costas et al., 2008).

Then, to solve the problem of testing engineering algorithms in real-time, we apply the advantages of the computer Network, computer communication and teleoperation. Furthermore, developing these new tools give the possibility to use these equipments for remote education.

In remote experimentation there exists several schemes based on the communication channel called **telepresence schemes**, some of them are: i) **teleoperation**, ii) **teleprogramming** and iii) **telecontrol**. In (Wang & James, 2005) some concepts are related with teleoperation. In other works, (Huijun et al., 2008) analyze the time-delay in the telecontrol systems, and (Cloosterman et al., 2009) studies the stability of the feedback systems with With Uncertain Time-Varying Delays. Others authors propose platforms only to move remote equipment, for example robots, (Wang & James, 2005). Finally, few works talking about the remote programming are published; see for instance (Costas et al., 2008).

However, for a remote laboratory to be functional, it must be capable of offering different schemes of telepresence. This can be easily understood from figure 1 which is an extension of the figure given in (Baccigalup et al., 2006). A comparison between different teaching methods, taking into account the teaching effectiveness, time and cost per students, is schematized in figure 1.

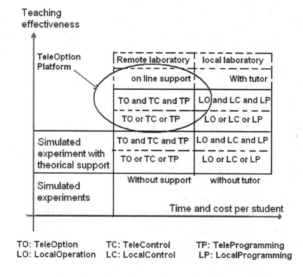

Fig. 1. Comparison between local and remote laboratories.

Contribution

Considering figure 1, the goal of this work is to introduce a platform called *Teleoptions*, which offers an alternative for remote laboratories, using three of the *telepresence schemes*: teleoperation, telecontrol and teleprogramming.

The main feature of this framework is its multitier architecture, which allows a good integration of both hardware (HW) and software (SW) elements.

Structure of the work

This work is organized as follows: In Section 2, definitions and concepts used in this work about tele-control, tele-operation and tele-programming are introduced. In Section 3, the proposed scheme based on multitier architecture is presented. The laboratory server description is given in Section 4. In Section 5, two applications of the platform are presented. The first application concerns the remote experimentation of an induction motor located in the IRCCyN laboratories in Nantes; France. The second application consists of the remote experimentation of the manipulator robot located in the CIIDIT-Mechatronic laboratories in Monterrey; Mexico. Finally, in Section 6, conclusions and recommendations are given.

2. Some concepts

Now, we introduce the concepts of teleoperation, telecontrol and teleprogramming, which will be used in the sequel.

Teleoperation is defined as the continuous, remote and direct operation of equipment (see figure 2). From the introduction of teleoperation technology, it made possible the development of interfaces capable of providing a satisfactory interaction between man and experimental equipment. On the order hand, the main aim of **telecontrol** is to extend the distance between controller devices and the equipment to the controller. Thanks to the development of the Internet, the distance between controller devices and the equipment has been increased (see figure 2).

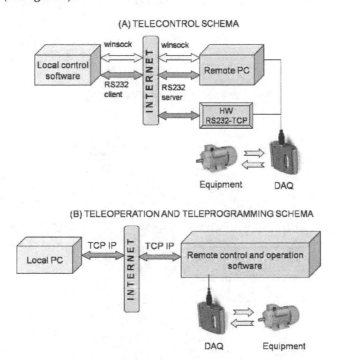

Fig. 2. Telecontrol, teleoperation and teleprogramming schema.

Figure 2.B shows a **teleoperation** scheme through the Internet working with a single channel of communication. This channel is used to change the parameters of the controller devices and/or plant. However, the effects of these changes will depend on the server layer.

Figure 2.A shows a **telecontrol** scheme through the Internet, in which the two channels of communications are required (closed-loop system), i.e. forward path *Ch1* and feedback path *Ch2*. In this case, it is necessary to maintain the stability of the closed-loop system. A solution to stability problem is that the time dalay must be less than the sampling period (Hyrun & Jong, 2005).

Furthermore, there exists a different interpretation about the **teleprogramming**. One of them is extending the distance between software programmer and the microcontroller or control board. On the other hand, it is possible to programming a remote system using two systems, called the master system and slave system, separated by the communication channel. In (Jiang et al., 2006) the teleprogramming method is based on teleoperation.

3. Framework proposed based on multitier programming

Now, we will introduce the software descriptions that are used in the proposed platform.

Figure 3 shows the tiers of the proposed framework called *Teleoption*, which has more performance than a classical telepresence framework application. *Teleoption* allows the interaction between different elements in hardware and software. Furthermore, it is possible to work under the three schemes of telepresence, i.e. teleoperation + telecontrol + teleprogramming.

The top level of the framework is the HTTP server, winsock services, webcam server and RS232 server. The second level of the framework implements the PHP script modules, DLL library and database services. All services can be shared by the VNC Server.

This distribution of software presents great advantages: i) Security in the platform, ii) several ways to transmit information from the hardware.

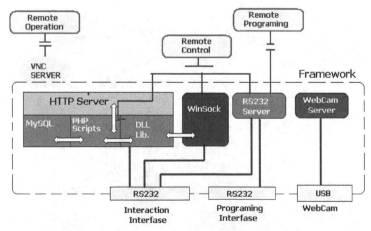

Fig. 3. Multitier architecture proposed.

Presentation tier. The HTTP Server is the presentation tier. This tier contains several Web pages with information of the platform services.

Furthermore it includes the instructions and regulation of the platform

Logic tier. In this tier, we have the programming layer. Three programming languages are used in the platform: PHP, Visual Basic and SQL. In the logic tier interacts the blocks: *i)* "PHP scripts" (which contain several programs in PHP) , *ii)* the block of the data base MySql and, *iii)* the block of the DLL libraries (designed in VBasic).

Database tier. The database tier contains information about of the platform, i.e. the users list, logbook. In fact, logic tier and database tier provide security to platform, since it is possible to use restrictions proportioned by a PHP script. This script allows the use of the platform only if the user has the permission.

Communication tier. The platform allow establish several ways of communication with the hardware: i) using *Serial Server Component* (RS232 Server), ii) using Windows sockets (Winsock) or DLL's library, and iii) using the PHP script services (see figure 4).

Serial Server Component is a software based RS232 to TCP/IP converter. RS232 Server allows any of the RS232 serial ports on the PC laboratory to interface directly to a TCP/IP network.

On the order hand, also is possible the remote access using the sockets of Windows or DLL's library. The remote user uses its own programs to send instructions to program modules of the platform.

Finally, the platform has modules designed in PHP, here, the remote user can to access to hardware using a Web page of the platform.

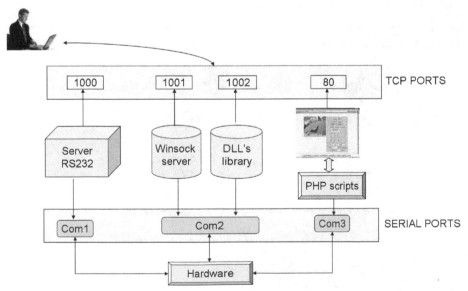

Fig. 4. Communication tier.

3.1 Operational method of the platform

When the services of **remote programming** are used, then the framework opens a communication's channel in order to share the serial services (RS232), and allows the remote programming.

If the services of **remote control** are used, then the framework opens more communication options. The first option is similar to the remote programming method, but in this case the control board and the equipment are separated, a remote communication is established by means of Internet using the services of the RS232 Server/Client.

The second alternative of remote control is the winsock option, which is similar to the last method, but the interchange of information is given by the winsock module. In this case, it is necessary to know the operation commands of the controller in order to send the information through that Internet to Winsock module, and then Winsock module will send the information to hardware.

The third option of remote control, the framework allows the access to control of the hardware using a Webpage, where the user does the work of controller. Here, the framework receives the commands of the user and sends this information to some PHP script, which sends the information to the operational layer of the multitier programming.

Finally, in the *remote operation*, all framework are shared using the services of some VNC (Virtual Network Computer) which is a communication protocol based on RFB protocol which allows the remote access of the desktop of other computers located on the web. VNC protocol transmits the keyboard and mouse events from one computer to another, relaying the graphical screen updates back in the other direction, over a network.

4. Laboratory Server (LS) implementation

Besides the proposed framework, an architecture based on *Computers of Distributive Tasks (CDT)* is proposed. This architecture is shown in figure 5.

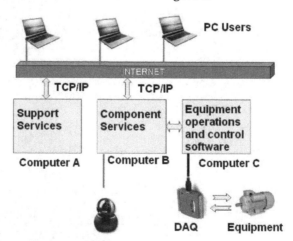

Fig. 5. Computers of Distributive Task.

Computer A allows establishing a communication both textual and oral between the local and remote user, in such way, this computer provides help on line and uses the following freeware software:

- Messenger: Textual communication and webcam.
- Skype: Oral communication, IP Telephony and videoconference.

Computer B has the task of sharing several resources through the Internet. The architecture proposed is installed in this computer. This computer uses the following software:

- Matlab/Simulink. This Software is used typically in control systems.
- ControlDesk. It is a graphical tool for controlling in real-time the equipment.
- UltraVNC server. It is software belonging to the VNC family
- LogmeIN. It is ESS software.
- TCPComm server. It is a RS232 server, which allows sharing the serial ports (COMM) of the computer. Serial port is used commonly as communication channel between PC and equipments.
- WebcamXP. Allow sharing the images from the webcams, these webcams can show the equipment details.

Computer C has an interface with the data acquisition board (DAQ), and does not share any resources on the Web. This computer is only used to share information with Computer B throughout the remote control. Furthermore, this computer protects the access to the plant (experimental equipment) in order to avoid damages caused by unauthorized users.

5. Experimental setup: Study cases

5.1 Remote experimentation of an electrical machine

The methodology described in the above section is applied to show remote access to the set-up of electrical motor located in the IRCCyN laboratory in Nantes France (figure 6), from the CIIDIT-Mechatronic laboratory in Monterrey, Mexico.

The set-up located at IRCCyN is composed of an induction motor, a synchronous motor, inverters, a real time controller board of dSPACE DS1103 and interfaces which allow to measure the position, the angular speed, the currents, the voltages and the torque between the tested machine and the synchronous motor. The motor used in the experiments has the following values: 1.5 kW normal rate power; 1430 rpm nominal angular speed; 220V nominal voltage; 7.5A nominal current; np = 2 number of pole pairs, with the motor nominal parameters: Rs = 1.633 Ohms stator resistance; Rr = 0.93 Ohms rotor resistance; Ls = 0.142H stator self-inductance; Lr = 0.076H rotor self-inductance; Msr = 0.099H mutual inductance; J = 0.0111/rad/s2 inertia (motor and load); fv = 0.0018Nm/rad/s viscous damping coefficient. The experimental sampling time T is equal to 200 s.

Furthermore, this laboratory is equipped with the remote technology described above, and can present several time delays that can appear during any real time experiments and are necessary to analyze:

- Transmission delay thought Internet (TI).
- Control algorithm computation (TC).
- Sampled time of the Data Acquisition (TS).

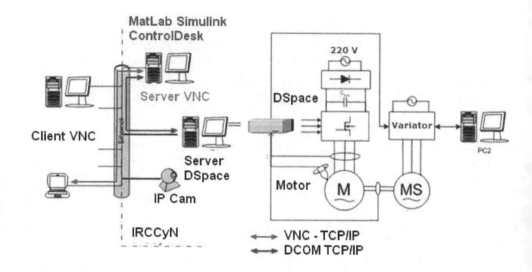

Fig. 6. IRCCyN laboratory schema.

These time delays depend on the tele-presence scheme selected. In a telecontrol scheme, the total time T = TI + TC + TS could be high and could affect the stability of the system. Nevertheless, if T = TC + TS is small, then a teleoperation scheme offers an excellent solution in remote experimentation, due to the time delay TI is not considered by the aforementioned reasons (see section 2).

Therefore, the scheme used for remote experimentation is based on teleoperation where the effects of the time delay and uncertain property is not considered in the stability of the system, because the controller and the plant are in the same layer, as shown in figure 2.

In this experiment, the time delays registered are: TI (ping) = 400 mseg. avg., TI (camera) = 3 seg. avg., TI (screen feedback, VNC) = 2 seg. avg., TC < 70 mseg.; TS = 120 seg. (DS1104).

Figure 7 shows a Mexican user, which applies a control algorithm, in order to access the remote laboratory, located in Nantes; France. From the figure 7, we can see *the computer A* showing the images sent by the webcam and the response obtained when the control algorithm is applied to the induction motor, which is transmitted by computer B using Controldesk and Matlab.

Figure 8 and 9 shows the screenshots obtained from this experiment. The first image shows the images given by webcam of the machine (with the sound), the second figure shows the Remote software ControlDesk throughout LogmeIn services.

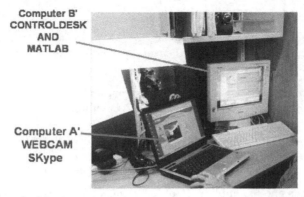

Fig. 7. Remote access by Mexican user.

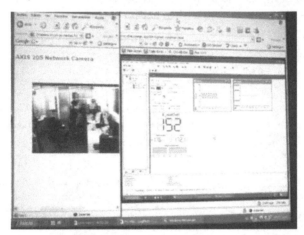

Fig. 8. Remote experimentation using LogmeIn services.

Fig. 9. Remote images of the induction motor.

5.2 Platform-setup in robotic education

It is undisputed that remote laboratories are not able to replace traditional face-to-face laboratory lessons, but they present some benefits of remote accessible experimentation:

- Flexible schedule vs. restricted schedule.
- Individual experimentation vs. group experimentation.
- Access from any computer vs. access only in the laboratory.
- Student self-learning is promoted.
- Student can use other educative means as Internet documentation, simulations, software, etc.
- The student is motivated when he is seeing his experiments and results.

This section presents another application of the architecture proposed. We emphasize that this architecture allows a remote user to access the services of control, programming and operation of robots located in the CIIDIT-Mechatronic laboratories in Monterrey; Mexico.

Teleprogramming. The objective of the teleprogramming is that the students use the BASIC microcontroller language in order to program the PICAXE microcontroller. In this platform, the student can use the basic instructions in order to program the robot: servo, *goto, serin, serout, pause, if, for*.

The student can program the PICAXE microcontroller using the flowchart method programming. Flowchart is an excellent means of pedagogy; the software shows a panoramic and graphical view of the programming sequence.

Telecontrol. The platform allows sharing the DLL resources so that the student can design programs in Visual basic, C, Matlab, or other languages. In the telecontrol option, the student can design and prove algorithms, using simulation software in local mode, subsequently if the capacity of the network is not large and it does not affect the stability of the systems, then it can be proven on-line on the robot.

Teleoperation. This platform offers the teleoperation services, so that the student can use all the services of the platform in remote mode. In this case, the platform shared the services of teleoperation using the Skype and logmeIn services.

Figure 10 showing the laboratory scheme located in CIIDIT laboratory in Mexico. The hexapod robot is acceded from the *PC Controller* Computer using two communication channels, RS232 and video. In the *PC Controller* Computer one is located the *Controller Module Server (CMS)*. The end user uses the services of the CMS in remote mode in order to control the hexapod robot.

Figure 11 showing the *screenshot* of a computed located in the IRCCyN Laboratory accessing to CIDDIT laboratory using the *LogmeIn* services.

- Figure 11.A shows the surroundings of the hexapod robot from a internal camera (eye hexapod).
- Figure 11.B presents the hexapod robot from a external camera (auxiliary camera).
- Figure 11 C shows the computers of the remote laboratory.
- Figure 11 D. showing Controller Module Client (CMC).

In the experiment, such a move-and-wait strategy is implemented of initiating control move then waiting to see the response of distant robot: then initiating a corrective move and waiting again to realize the delayed response of the distant system and the cycle repeats until the task is accomplished.

Let us define N(I) to be the number of individual moves initiated by the operator according to the move-and-wait strategy. The number N(I) depends only on the task difficulty and is independent of the delay value according to experiments (Hocayen & Spong, 2006). Consequently, the completion time, t(I), of the certain task can be calculated based on the value N(I) as follows:

$$t(I) = t_r + \sum_{i=1}^{N(I)} (t_{mi} + t_{wi}) + (t_r + t_d)N(I) + t_g + t_d \tag{1}$$

Where $t_r, t_{mi}, t_{wi}, t_g, t_d$ are human`s reaction time, movement times, waiting times after each move, grasping time and delay time introduced into communication channel, respectively.

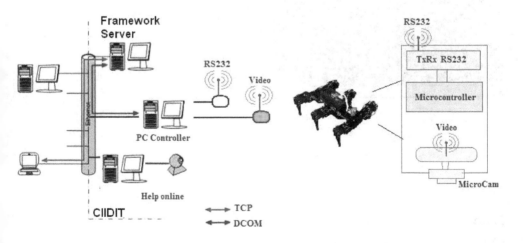

Fig. 10. CIIDIT Laboratory schema.

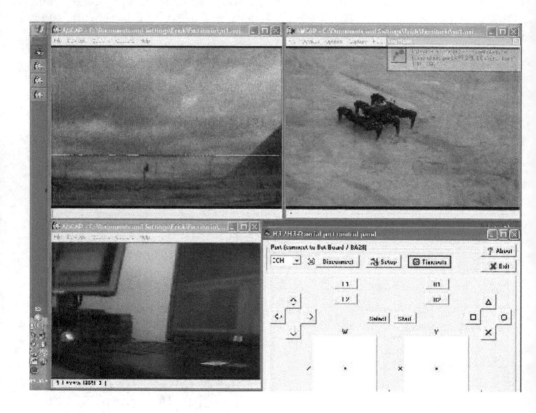

Fig. 11. Experimentation from IRCCyN, Nantes France.

6. Conclusions

In this work the capability of interfacing a large set of options with remote experimentation through the Internet has been demonstrated by the architecture based on multitier architecture.

This architecture allows the easy integration of both hardware and software, offering an excellent tool for remote experimentation, which allows the experimentation using the teleoperation, the telecontrol and teleprogramming schemes.

The main characteristic of the proposed platform has been outlined in this paper by means of a description of experiments.

7. Acknowledment

This work was supported by CONACYT, ECOS-NORD, PAICYT-UANL, Mexico and France.

8. References

Baccigalup, A.; De Capua, C.; Liccardo, A. (2006) Overview on Development of Remote Teaching Laboratories: from LabVIEW to Web Services, Instrumentation and Measurement Technology Conference, Sorrento, Italy, pp. 24-27.

Callaghan, M. J.; Harking, J.; El Gueddari, M.; McGinnity, ATM; Magure LP (2005) Client-Server Architecture for Collaborative Remote Experimentation, Procedings of the ICITA 2005, 0-7695-2316-1/05 IEEE.

Cloosterman, M.B.G.; van de Wouw, N. (2009); Heemels, W.P.M.H.; Nijmeijer, H.; Stability of Networked Control System with Uncertain Time-Variing Delay. Automatic Control, IEEE Transactions on, Volume 54, Issue 7, pp. 1575-1580.

Costas-Perez, L.; Lago, D.; Farina, J.; Rodriguez-Andina, J. (2008). Optimization o fan Industrial Sensor and Data Acquisition Laboratory Through Time Sharing and Remote Access. Industrial Electronics, IEE Transactions on, Volume 55, Issue 6, pp. 0278-0046.

Davoli, Franco; Spano, Giuseppe; Vignola, Stefano; Zappatore, Sandro. (2006) Towards Remote Laboratories With Unified Access, IEEE Transactions on Instrumentation and Measurement", Vol 55, No. 5.

Gomez, Luís; Garcia, Javier (2007); Advances on Remote Laboratories and e-learning experencies. Deusto Publicaciones, ISSB 975-84-9830-662-0

Hokayen, Peter F.; Spong, Mark W. (2006) Bilateral teleoperation: An historical survey, Automatica 42 : 2035-2057

Huijun Gao; Tomgwen Chen; James Lam (2008); A new delay system approach to network-based control. Automatica, Volume 44; Issie 1, pp. 39-52

Hyun, Chul Cho; Jong, Hyeon Parck (2005) Stable bilateral teleoperation under time delay using robust impedance control. Mechatronic, Vol. 15: 611-625.

Jiang, Zainan; Xie, Zong; Wang, Bin; Wang, Jie; Liu, Hong (2006) A teleprogramming Methos for Interned-based Teleoperation. International Conference on Robotics and Biomimetics, Dec. 17-20, Kuynming China.

Rapuano, Sergio; Zoino, Francesco (2005) A learning Management System Including Laboratory Experiments on Measurement Instrumentation, IMTC 2005, Instrumentation and Measurement Technology Conference, Ottawa, Canada, pp. 17 - 19 .

Restivo, M.T.; Mendes, J.; Lopes, A.M.; Silva, C.M.; Chouzal, F (2009). A Remote Laboratory in Enginnering Measurement. Industrial Electronics, IEEE Transactions on. Volume 56, Issue 12, pp. 4836-4843.

Wang, Meng; James N.K (2005) Interactive Control for Internet-based Mobile Robot Teleoperation, Robotics and Autonomous System 52, pp. 160-179.

Wu, Y. L; Chan, T.; Jong B.S.; Lin, T.W. (2008) A Web-based virtual reality physic laboratory", In Pro 3rd IEEE ICALT, Athenas Grerce, pp.455.

Part 3

Sensor Networks

7

Power Considerations for Sensor Networks

Khadija Stewart[1] and James L. Stewart[2]
[1]*DePauw Universtity*
[2]*Purdue University*
USA

1. Introduction

Wireless sensor networks (WSNs) are networks composed of small, resource-constrained and collaborative devices. WSNs are used in a plethora of domains including environmental and agricultural monitoring, military operations, in the health care field and in building automation. The three main functions of wireless sensor nodes (also called motes) are to sense the environment, perform computations, store intermediate results and communicate with other motes in the network.

This chapter focuses on power considerations for all aspects of wireless sensor networks. It covers software, hardware and networking aspects of the motes. The main limitation of wireless sensor motes is that they operate on battery power. In many WSN applications, the motes are placed in remote areas and deployed for the lifetime of the network. During this time the only power resource the motes have access to is their battery. An example of such a deployment is the Mount St. Helens project developed to study volcanic activities on Mount St. Helens (were volcanic eruptions can occur at any time with very little warning). The sensors were placed on the mountain using helicopters and work at length to continually sense seismic activity and relay information to a data center. For such applications, the battery lifetime is the main factor that dictates the lifetime of the network. It is therefore imperative to develop wireless sensor mote platforms that minimize the power consumption and/or maximize the lifetime of the network as a whole.

Several works in the literature address one or two aspects of the mote's architecture and/or functionality but to the authors' knowledge, no work has combined all said aspects and addressed them as a homogeneous unit. This chapter studies and analyzes each hardware component of the mote's architecture, all the main protocols used in the mote's stack layer, discusses the work that has been done in terms of reducing the power consumption, increasing the battery lifetime and or increasing the lifetime of the entire network as a whole.

The chapter is organized as follows: Section 2 gives an overview of wireless sensor networks, their applications and general architecture. Section 3 focuses on the hardware architecture of the motes (the CPU, communication infrastructure, memory and sensors). Section 4 introduces the layered protocol stack of the sensor motes (application, transport, network, link and physical layers). Section 5 summarizes the chapter and suggest paths forward.

2. Preliminaries

Wireless sensor networks are composed of small, inexpensive devices that are designed to sense some phenomena, perform light computations and communicate with one another. These devices are usually scattered over some area. This technology has seen a wide range of applications ranging from military use to personal security. In the following, we discuss the history of WSNs and some of their most pertinent applications.

Wireless sensor networks evolved form the Smartdust project, which was developed and funded by DARPA in the late 1990s. The Smartdust project was designed to show that "a complete sensor/communication system can be integrated into a cubic millimeter package" (Pister, 2001). The Smartdust motes were engineered to be power efficient. This and other similar projects have led to the explosion of research in the area of wireless Ad Hoc and sensor networks, which was and still is heavily supported by US government agencies including the National Science Foundation. While working on the Smartdust project, the researchers recognized the variety of applications for their work both in the military field and elsewhere.

Some of the applications for the Smartdust projects are virtual keyboard, inventory control, product quality monitoring and smart office spaces among others (Pister, 2001). In the virtual keyboard application, dust motes would be glued into fingernails to sense the orientation and motion of the fingertips and communicate with a computer. This could be used in sign language translations, piano play, etc... In the inventory and quality control applications, a system of communication could be implemented and deployed in all aspects of the production process in order to monitor the location of the product and control and monitor its quality (from temperature, to humidity exposure etc...). In the smart office spaces application, the person's preferred temperature, humidity settings could be directly communicated to the environment they walk into. Some of the military applications that the Smartdust project was developed for include battlefield surveillance, transportation monitoring and missile monitoring.

In the past few year, Wireless Sensor Networks made the transition from the Berkeley research centers to the production arena with the creation of companies, such as Crossbow Technologies (Crossbow Technologies, n.d.) that started manufacturing them. The appeal of Wireless Sensor Networks stems from the fact that you can deploy them and just leave. We discuss in the remainder of this section the main classes of applications for the general WSNs.

WSN applications can be categorized into habitat and environmental monitoring, heath applications, commercial applications, military applications among others.

One of the most prevalent uses of WSNs is in habitat and environmental monitoring. It has been shown that direct human contact with some plant or animal colonies can result in disastrous consequences. For example, (Mainwaring et al., 2002) describe the use of a sensor network to monitor Seabird colonies because of their sensitivity to human disturbance. In fact, a 15 minute visit to the colony could result in up to 20% rate of mortality among eggs. Not only are WSNs useful in monitoring colonies without causing any disturbances but they also represent a more economic method of monitoring for long periods of time.

Another example of the environmental use of WSNs is in forecasting systems. WSNs are now scattered around large areas to forecast pollution, flooding and seismic activity. The Automated Local Evaluation in Real Time (ALERT) was developed in the 70s by the National Weather Service. It has been used by several government and state agencies and international organizations to provide a real time data collection system that can forecast floods (ALERT, n.d.).

Another use for WSNs is in intelligent building management. In fact, they have been used in HVAC, lighting, climate control, fire protection, energy monitoring and security applications among others. In Canada for example, the National Research Council launched a three-year project to develop wireless sensor networks to do just that. The project started in 2008 and is anticipated to continue through 2011.

A very important application of WSNs is in the healthcare field. WSNs can be used to provide continuous, remote, inexpensive, instantaneous and non-invasive monitoring of a patient's vital signs. This technology can be used to allow the elderly to remain in their own residences but still be able to continuously check their vitals.

All these WSN applications consist of deploying the network for an extended period of time on a single battery charge. It is therefore imperative that the motes be power efficient and that the lifetime of the network as a whole be as long as possible.

3. The WSN hardware architecture

The hardware of wireless sensor motes consists of sensors (analog and/or digital), a microcontroller, also referred to as a microprocessor or Central Processing Unit(CPU), memory, RF communication module (transceiver) and battery. The design of each component of a WSN mote should take into consideration the power metrics (power consumption and voltage requirements) of the component. Additionally, the integration/interface of all the components as a whole should be studied for power consumption (having analog sensors means that an ADC component in the CPU should be required to convert the sensor readings to a digital format etcÉ)

To reduce power consumption, several works suggest the introduction of sleep and wake up cycles for the motes. Other schemes suggest a better integration of the functionality of hardware components (using cross-layer principles). Another consideration in the design of the CPU is the clock component. Several applications of WSNs require some level of time synchronization. Clock choices and designs affect the amount of drift that a sensor mote's clock can experience requiring more or less time synchronization operations when the mote is deployed (Akyildiz et al., 2002).

3.1 Sensors

Of the five main units, the sensing unit is the most application specific. Meaning the type of sensor used will depend on the application. For instance, wireless sensors used for structural health monitoring may consist of materials apt for monitoring strain, acceleration (accelerometer) and linear and angular displacement. Other application specific sensors may measure, vehicular movement, soil consistency, blood alcohol levels, humidity, noise levels and so on. These sensors should then report some signal indicative of their acquisition. Temperature (thermo-coupler outputting voltage or thermistor outputting resistance), force or pressure (piezoelectric outputting voltage, strain gauge outputting resistance), position (linear variable differential transformers (LVDT) outputting alternating current) or light intensity (photodiode outputting current) all need to report information regarding their surroundings to a processing unit (Wilson, 2004).

The development of an efficient method for acquiring and converting conventional energy from the sensors such as solar and wind has seen an exponential growth over the last few years. The sensing development has been referred to as energy harvesting. One factor contributing to the enjoyment of such an increase has been the threat of rapid decreases projected in our global and national energy reserves based on utilization rates and trends. Such a premonition has spurred funding for research in various fields including materials or more specifically metamaterials.

Metamaterials have been defined by most associated scientist as materials made by man which exhibit non-natural properties and characteristics, particularly EM or electromagnetic properties not known to exist with any other materials found in nature. Regarding electromagnetism, metamaterials which exhibit propagating electromagnetic waves (both the permittivity and permeability are negative as seen in Figure 1 have seen much attention in recent years as well as when both permittivity and permeability are very close to 1.

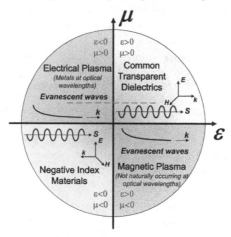

Fig. 1. The parameter space for ϵ and μ. The two axes correspond to the real parts of permittivity and permeability, respectively. The dashed green line represents non-magnetic materials with $\mu = 1$ Cai &shaleav (2010).

The reaction or response of a material (as in a sensor for WSNs) to external fields is largely determined by only the two material parameters ϵ and μ, permittivity and permeability respectively. As shown in Figure 2, the real part of permittivity ϵ_r is plotted to the horizontal axis of the parameter space, while the vertical axis corresponds to the real part of permeability μ_r. A negative value of ϵ (μ) indicates that the direction of the electric (magnetic) field induced inside the material is in the opposite direction to the inbound incident field. Noble metals at optical frequencies, for example, are materials with negative ϵ, and negative μ and can be found in ferromagnetic media near a resonance. Waves can not propagate in material in the second and fourth quadrants, where one of the two parameters is negative and the index of refraction becomes purely imaginary. In the domain of optics, all conventional materials are confined to an extremely narrow zone around a horizontal line at $\mu = 1$ in the space, as represented by the dashed line in Figure 2.

Scores of such materials are designed to manipulate EM waves, many passively, by creating an alternate propagation path. Metamaterials have been designed to redirect, not absorb or

Fig. 2. Examples of a 3D optical metamaterials fabricated in a layer-by-layer manner. (a) A near-infrared NIM (Negative Index Material) with three functional layers made by EBL (Electron Beam Lithography); (b) Four layers of SRRs (Split Ring Resonators) based on EBL with patterning-and-flattening approach; (c) A NIM wedge exhibiting negative refraction for visible light made by an advanced FIB technique Cai &shaleav (2010).

reflect but to divert the energy through a desired path. It is no wonder as to the attention metamaterials have seen for energy harvesting. Research is currently being conducted to develop sensors (both photon and electron based) that extract atmospheric energy regardless of the incident angle such that no energy is reflected back out of the sensor rather, it's reflected down toward the detector. This will lead to the creation of ultra-efficient sensors for wireless networks, see (Narimanov & Kildishev, 2009; Shalaev et al., 2005) for more information on Metamaterials.

3.2 Microcontroller

The component responsible for doing the bulk of the switching and decision making for the WSN at the remote site is the processor or microcontroller. When selecting the processor for specific WSN applications, the engineer must make many considerations. These considerations include, but are not limited to, cost, power requirements, physical size, weight and speed, some of which are elaborated upon below.

Depending on the microcontroller, the power requirement could range from .25 mA to 2.5 mA per MHz for either 8 or 16 bit processors. Again, the application will determine if a processor consuming relatively high amounts of energy is acceptable or if .25 mA per MHz is needed. A common misconception is that by putting the processor in "sleep" mode, the sensor utilizes less power thus is more efficient. It has been shown that this is not always true as while in "sleep mode", sensors still maintain synchronization and memory functionalities necessary to perform expeditiously upon awakening (Hu & Cao, 2010).

In fact a more prudent approach to saving energy would include completely shutting the processor off, entirely, and ensuring the sensor can rapidly recover from a "dead" state or at the very least rapidly jump from "sleep" mode to "awake" mode. As the processor needs to synchronize native clocks and stabilize, the transition time or delay can be as long as 10 ms which is a relative eternity. Another parallel approach involves varying the speed depending on the time allotted for a specific task.

In other words, only using the minimum power required for a task at a given time by dynamically ramping up or down the power accordingly versus drawing full power for all "awake" states. This approach may benefit from an algorithm in which the speed is a function of the power. If the required task and itÕs effort expended is known before the task is given, an absolute "finish time" can be maintained without necessarily completing the task as fast a possible rather as fast as necessary. Researchers from the University of California, Irvine (Irani et al., 2007) developed an algorithm for optimizing power consumption by varying speed below:

$$g(z,z') = \frac{\sum_j \, such \, that \, [r_j, dj] \subseteq [z,z'] R_j}{l(z,z')} \tag{1}$$

where g(z,z') defines the intensity of the interval [z,z'], l[z,z'] defines the length of the interval, R_j is the required work needed to complete the job and d_j denotes the deadline for job j. This would allow energy and speed to be spent where it's needed most creating a dynamic fluid speed variance throughout the CPU for maximum overall efficiency. One might say, 'losing a battle here and there but winning the war'.

3.3 Memory

Memory is a crucial factor in WSNs. Particularly non-volatile memory. Non- volatile memory is defined as various forms of solid state memory which doesn't need to be refreshed or powered to maintain it's information. Examples include flash, electrically erasable programmable read-only memory (PROM) read only memory (ROM), optical discs and magnetic disks (Postolache et al., 2010).

The memory component is the means at which the WSN stores the data it acquires. The speed requirement of the memory unit depends of the nature of the WSN and its intended functionality. A rather fast memory unit may be required for certain military applications where the data acquisition speed from the memory may dictate whether or not a target is detected in time for acquisition and lock. On the other hand, a relatively slow memory unit may be acceptable for soil monitoring WSN utilized by farmers. In either case the security and reliability of the memory unit is important and both require additional power demands on the WSN. To this end, researchers have been developing ways to more efficiently processing and storing the acquired data including virtual memory protocols. Virtual memory has been shown to reduce compile-time optimizations regardless of the limitations in memory on site. One approach which generates a memory layout optimizes to the memory access patterns and attributes for a given WSN. In other words, the protocol optimizes the memory map based on the application, effectively reducing overhead (Lachenmann et al., 2007).

3.4 Transceiver module

All WSN motes will possess a transceiver or TR modules as they allow the motes to communicate in WSNs. They present the capability to transmit and receive data packets, information or signals in a relatively small package. One of the main factors which allows for such a diminutive size lies in the RF architecture. Because the TR modules transmit and receive in the same RF component there is no need for a separate architecture for each transmission or reception. Thus the isolation of incident energy to transmitted energy must be great to ensure against destructive cross modulation, unwanted dispersion and various other resultant noise, all of which would inherently degrade the efficiency of the WSN either directly or indirectly. Signal loss is of particular concern in the input/output portion of the TR module and precautions must be taken to ensure signal degradation is tolerable from a minimum threshold point of view.

Within the TR package, a typical TR module will consist of and follow this RF path for transmission: a common attenuator for signal suppression, a common phase shifter (depending on the application. For example, phase shifter could be used to shape the transmission pattern or radiation pattern leaving a WSN (also known as beam-steering), a driver and a high power amplifier (HPA) to boost the signal amplitude for propagation from the aperture or antenna for transmission. When receiving a signal within the TR module frequency range, which varies per application, the signal passes through a limiting filter and low noise amplifier (LNA) before coursing through a common attenuator to suppress the signal's magnitude and possibly a common phase shifter (depending on the application. For example the phase shifter can be utilized as a directional finder or filter for incident signal in a WSN). Outbound and incident signals are typically discerned by a circulator at the output/input of the module. The attenuator and phase shifters are termed "common" due to the fact that these components are used for both reception and transmission. In the following, we elaborate on a few of the key components of the TR module from Figure 3.

The attenuator is implemented to ensure the unwanted side-lobes are suppressed, sufficiently reducing the noise in the system. It also keeps the amplifiers down stream from prematurely reaching saturation and causing unwanted non-linearities. Typically this is done only for the receiver as during transmissions, it is usually desirable to propagate as much energy from the antenna as possible. Since the attenuator basically performs the exact opposite function of the amplifier, they are typically not conjoined in series unless, in some cases, it's needed for filtering purposes. Note that all the components within the TR module are frequency matched meaning they are optimized for specific frequency ranges. Due to this inherent characteristic, attenuators can be used to suppress frequency bands without distorting the fundamental waveform. This is important for the energy efficiency of the system as the modulator can maintain relative simplicity without the need to effectively recreate a waveform which would subsequently cost more power.

The phase shifter allows multiple RF signals to be controlled by way of an external stimulation such that the output of the phase shifter is of the desired phase without effecting the frequency. The phase shifter may or may not be present in the TR module. It depends on whether or not the WSN calls for a beam-forming or shaping capability which can aid in power efficiency if multiple sensors are synchronized in receive and or transmission mode for power/amplitude coupling. The amplifiers (driver and high power amplifiers) boost the signal for transmission from the antenna. The level of amplification needed depends on the efficiency of the system, particularly the aperture or antenna. A poorly matched antenna or one which has a high Voltage Standing Wave Ratio (VSWR) will demand a higher amplitude or stronger signal to propagate to a given target.

The application and placement availability of WSN will greatly affect which antenna is more suitable and efficient. Most WSN antennas are omni-directional fundamentally but are shaped by various ground effects. This crucial aspect of antenna propagation has prompted many researchers to develop accurate prediction models specifically for WSNs.

3.5 Power source

Considering that many WSNs rely on portable energy or power sources to power sensors, the capacity and efficiency of both the power source and the WSN is crucial in the overall effectiveness of the WSN. For most of the WSN applications, when the power source drains,

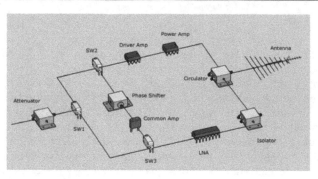

Fig. 3. Transceiver

the WSN is inoperable. For many applications various protocols for maximizing the lifetime of the WSN are adequate while many other applications require WSNs to remain in remote areas for several months or years without opportunities for manual power replenishment. Many research centers have developed models to efficiently harvest energy for power as for sensing previously mentioned. A WSN which can obtain its power requirements from its surrounding environment essentially has an infinite lifetime. Various approaches from mechanical vibrational energy harvesting to photon collection schemes are being considered in an effort to self-generate power needs.

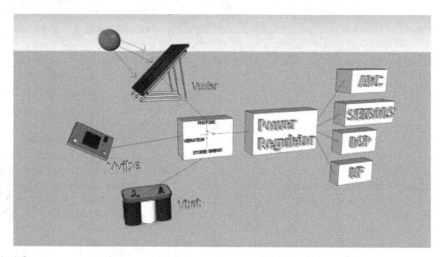

Fig. 4. A low power wireless sensor node system powered from energy scavengers or harvesters and a battery. Guilar et al. (2006).

Figure 4 is a schematic of a low power WSN system that uses energy scavengers. In Figure 4, the energy sources are labeled Vsolar, Vvibe and Vbat for the solar, mechanical vibration and battery, respectively. A mutiplexer switches between the unregulated energy sources. ADC denotes the Analog to digital converter, DSP denotes the Digital signal processors and RF denotes Radio frequency.

4. The WSN layered protocol stack

The WSN layered protocol stack consists of the Application layer, the Transport layer, the Data link layer and the Physical layer. This section will cover the role of each layer and study its power consumption. The section will survey the current literature and analyze it with respect to power consumption.

4.1 Application layer

The application layer is in charge of collecting and processing sensor readings (including the use of data aggregation), performing time synchronization, implementing a security protocol (as needed) etc... Each one of these tasks uses one or more hardware modules and each task results in power being consumed.

4.1.1 Information fusion

Traditionally, sensor motes were designed to perform very little to no processing. They would sense the environment and send the sensing data to the base station for processing. This resulted in large amounts of packets being sent from the motes to the base station. In addition, in several sensor network applications, the motes are exposed to conditions (such as very high/low temperatures, pressure and noise) that might sabotage their measurements. It was then proposed to use information fusion (also referred to as data aggregation) techniques at the motes in order to decrease the network traffic, save energy, remove outlier data, make predictions about future measurements and in general obtain better information quality by combining data from multiple sources. Data aggregation requires some amount of processing to be carried out at the motes. Data fusion can be used at different layers of the WSN protocol stack. For example, it can be used at the application layer to process sensor readings as well as at the network layer to consolidate routing information. In the following, we survey and analyze the work that has been done on data aggregation and information fusion.

Information fusion can be categorized into three classes. Complementary, redundant, and cooperative. This classification depends on the particular application and the relationship between the motes that gather the data. In the case of complementary information, sources gather different types of data and information fusion is applied to obtain a more complete picture from data. In the case of redundant information, one or more sources gather the same type of data and information fusion is used to discard the outlier measurements and filter the data for accuracy, reliability and confidence. In cooperative information fusion, two sources gather information that is fused to produce information that better represents the reality. Information fusion is performed for different purposes. In the following, we present a classification of data fusion algorithms based on the purpose of the information fusion.

Information fusion techniques could be either centralized or distributed. Centralized techniques have a single point that controls the fusion process but are simple to implement. However, all the sensor motes send their data to the central point, which overwhelms the central data point and floods the network with messages. Distributed techniques on the other hand are more complex to implement but are more energy efficient because the information is exchanged locally, which reduces the number of messages exchanged in the network. Several methods have been proposed for information fusion including inference, estimation, aggregation, and compression.

Inference methods consist of making a decision based on previous knowledge. Protocols that have been proposed for inference include Bayesian inference algorithms (Coue et al., 2002; Tsymbal et al., 2003), Dempster-Shafer inference algorithms (Dempster, 1968; Shafer, 1976) and fussy logic algorithms (Gupta et al., 2005) among others.

Estimation methods use probabilistic theory to estimate a state based on a sequence of measurements. Estimation algorithms include the maximum likelihood algorithm (Xiao et al., 2005), least square, Kalman filter and particle filter (Kalman, 1960).

Data aggregation methods are used to overcome implosion and overlap and compression is used to reduce the amount of data by exploiting spatial correlation among the motes. Techniques for compression include distributed source coding (Xiong et al., 2004) and coding by ordering (Petriovic et al., 2003).

The implementation of these algorithms comes at a cost involving hardware complexity, CPU time and energy.

4.1.2 Time synchronization

Time synchronization consists of synchronizing the local clocks of all the members of a distributed network. In WSNs, it consists of synchronizing the clocks of all the motes in the network. Time synchronization is essential for all networked systems and is a requirement in most WSN applications and protocols. Example applications include environmental monitoring and target tracking among others. In these applications, the order of events is usually important. For example, in target tracking, sensors need to continuously report the location of a moving target, which could be time sensitive. Example protocols that require time synchronization are some MAC protocols (Demirkol et al., 2006) (such as the ones based on TDMA, where each node is assigned a time slot) in addition to several routing and security protocols.

In this section, we review the time synchronization algorithms proposed in the literature and analyze their power saving properties. In addition to being energy efficient, time synchronizations schemes for WSN need to be accurate and scalable.

When a packet is sent from node A to node B, node A can append a time stamp to the packet. Node B can then extract the time stamp from the packet, add the time it took the packet to travel from node A to node B (transmission time) in order to estimate its local clock's drift from node A. The packet delay consists of send time, access time, propagation time and receive time. Send time is the time interval between when the node issues the send command until the node is ready to send the packet.

The medium access times is the duration from when the node is ready to send until the time when the transmission starts. This is the step that makes time synchronization such a difficult problem. It is not possible to accurately estimate this time. The propagation time is the time it takes the packet to reach to the destination, and the receive time is the time it takes to receive the frame.

The Network Time Protocol (NTP) described in (NTP, n.d.), is the protocol that synchronizes the clocks in wired networked systems by estimating the roundtrip time of packets. It is the standard used on the Internet. NTP maintains a universal time (UTC) across the network. NTP is not suitable for WSNs because of its centralized nature and prohibitive cost. In fact in NTP, clients synchronize their clocks to the server and servers are synchronized to using external time sources (using a GPS). NTP is not suitable for WSNs for a number of reasons. First, NTP is centralized. 2. In WSNs, it is impossible to accurately estimate the roundtrip time. 3. GPS is too expensive to use or is not an option for most WSN applications (for example, indoor applications will not have access to GPS signal).

Most time synchronization protocols are sender to receiver. The sender time stamps a packet and the receiver extract the time stamp and tries to extrapolate its clock drift compared to the sender (Romer, 2001), (Ganeriwal et al., 2003). However, the Reference Broadcast Scheme (RBS) (Elson et al., 2002) is different. It is a receiver-to-receiver synchronization protocol. In RBS, a sender broadcasts a beacon without any time information. The receivers then exchange Acknowledgement messages with the time they received the beacon. Receivers can then extrapolate their own clock drift relative to each other. RBS works with two receivers and is easily extended to more than two receivers. In addition, increasing the number of broadcasts increases the accuracy of the scheme. Note that in RBS, the uncertainty of access time is removed (since the sender is removed from the drift calculations) and since the propagation time is assumed to be negligible in WSNs, the only uncertainty factor and potential error margin in this protocol is the receive time.

The Timing sync Protocol for Sensor Networks (TPSN) (Romer, 2001), (Ganeriwal et al., 2003) is a sender-receiver protocol. In TPSN, the sender sends a packet to a receiver, which uses the TPSN equation to extract its local clock drift compared to the sender. TPSN then uses a tree hierarchy to propagate the synchronization, it categorizes the nodes in the network into levels during the discovery phase. During the synchronization phase, the root node (level 0) synchronizes all level 1 nodes. After this first phase of synchronization, level 1 nodes synchronize level 2 nodes and so on, until synchronization has been propagated through the entire network. TPSN achieves better accuracy than RBS when using MAC layer time stamps because RBS is limited by the transmission range and would require more beacons in order to perform synchronization.

In (Greunen & Rabaey, 2003), the authors claim that most sensor network applications do not require very precise synchronization. In fact, they claim that most applications only require synchronization in the order of a fraction of a second. The authors therefore propose a different approach where the required accuracy is taken as a constraint and then a synchronization algorithm with minimal complexity is devised so that the requested accuracy can be achieved. In this work, the synchronization is propagated in a centralized manner where a spanning tree is created and synchronization is conducted along the edges of the tree.

Centralized approaches to time synchronization are not energy efficient and often result in depleting the energy reserves of the root node. The authors in (Maroti et al., 2004) propose the Flooding Time Synchronization Protocol (FTSP). FTSP uses periodic flooding of synchronization messages. This approach makes the algorithm de-centralized, scalable and topology independent. In FTSP, the synchronization root is elected dynamically and re-elected periodically. The root is responsible for keeping the global time of the network. In this work, the nodes form a dynamic mesh like structure to propagate the time synchronization throughout the network (unlike TPSN). This work saves on the energy required to create an initial spanning tree (as in TPSN) and is therefore more energy efficient than TPSN. In addition, this protocol is not topology dependent and can perform synchronization even when the topology of the network changes. However, the synchronization error in FTSP can grow exponentially with the size of the network (Lenzen et al., 2009).

Similar to FTP, the Novel Algorithm for Time Synchronization (NATS) is a decentralized time synchronization protocol. Unlike FTSP, NATS is a receiver-sender protocol because the receiver requests synchronization from the sender. This reduces the amount of messages exchanged for the purpose of time synchronization and therefore, reduces the amount of energy consumed during synchronization. NATS was designed at DePauw University by Peter Terlep [1] , Steven Klaback [2] and Khadija Stewart. NATS does not need to meet any specific topology prerequisites, it can adjust to topology changes. It accomplishes the following: 1) it does not need a third party device that is within radio communication range of all motes, 2) it does not need any one mote to be within range of all motes, 3) it is scalable, 4) it allows for deep sleep between synchronization activities, 5) it handles receiver-side medium access control (MAC) buffer latency uncertainty, 6) it addresses the inability to acquire a real-time sender-side MAC timestamp, 7) and it uses a distributed energy efficient algorithm for multi-hop synchronization.

Pair-wise synchronization in NATS starts when the root node receives a sync request. The root then sends two consecutive packets to the requesting node, each containing a timestamp at the MAC layer. The receiving node uses these two packets along with its receive time stamp to extrapolate the propagation and channel access times. It uses that information to estimate its clock drift from the root node. Time synchronization is then propagated throughout the network in a distributed manner, similar to FTSP, by having each synched node act as a potential root node for synchronization. Experimental results show that NATS provides better synchronization accuracy than TPSN. In fact, using the Sun Spots platform, the Mean Sync Error for NATS was 1.74ms versus 2.63ms for TPSN.

The Gradient Time Synchronization Protocol (GTSP) is completely distributed (Sommer & Wattenhofer, 2009), where the nodes periodically broadcast synchronization beacons to their neighbors and agree on a common clock. It is proven that after multiple beacon exchanges, the clock of the nodes converges to a common value. This algorithm is completely distributed and nodes only exchange beacons locally. GTSP is proven to achieve better time synchronization accuracy as compared to tree-based methods.

In PulseSync (Lenzen et al., 2009), the root node floods a "pulse" through the network in a breadth-first search tree manner. The nodes receiving the pulse then compensate for the drift relative to the root node. The authors note that the flooding of the pulse needs to be scheduled in order to avoid collisions. This protocol is proven to be accurate when used in sensor network applications where the topology does not change. In fact, it is proven to outperform FTSP by a factor of 5 on mid-size networks.

The authors in (Li et al., 2011), propose a new direction in time synchronization where the Radio Data System (RDS) of FM radios is used to synchronize the nodes' clocks. In this work, each node is equipped with an FM receiver and programmed to receive the same RDS signal. The mote's clock then uses a calibration component to calibrate itself to the RDS clock. The drawbacks of this method stem from the fact that the FM interface is not power efficient and

---

[1] Peter Terlep is currenty a Ph.D student at Michigan University
[2] Steven Klaback is currently with Digital Knowledge

that not all WSN applications can have access to FM signals especially for the applications deployed in remote areas.

In summary, in order for a time synchronization protocol to be appropriate for a wide range of WSN application, it needs to accurately compute the clock drifts, be distributed, scalable, adapt to any topology and be able to propagate the synchronization instantaneously and without flooding the network.

4.2 Transport layer

The transport layer is mainly used to communicate with external networks (such as the Internet) and is therefore rarely implemented in sensor motes.

4.3 Network layer

The network layer is in charge of all routing functions. Routing is the function that is used the most in multi-hop WSNs. It is the routing algorithm that allows nodes that are more than a hop away to communicate with each other and form a connected network. Because routing is used extensively in most WSN applications, it is the function that should be the most power efficient. A variety of routing protocols have been proposed in the literature, some of which are designed to be 'power aware' and use the battery level or the network lifetime as a routing constraint. This Section reviews these works and studies the effect of clustering on power consumption.

Initially, research on routing algorithms focused on Mobile Ad hoc NETworks (MANETs). In these networks, the nodes were designed to be highly mobile, which resulted in the development of on-demand routing algorithms. These algorithms use flooding to compute routes (see the Reliable Ad-Hoc On-Demand Distance Vector Routing Protocol (RAODV) (Khurana et al., 2006), and the Ad hoc On-demand Multipath Distance Vector (Marina & Das, 2001) among others). The traditional flooding method consists of every node broadcasting the data to all its neighbors, the neighbors broadcasting the data to their neighbors etc... Ultimately, the sink will overhear the data. Flooding-based protocols suffer from several inefficiencies including overwhelming the network with unnecessary transmissions, excessive energy consumption, implosion, overlap, among others see (Heinzelman et al., 1999). Routing in MANETs is a tedious problem because of their dynamic nature. Adding power efficiency to the equation renders the problem even more tedious.

In (Mleki et al., 2002), the authors propose a reactive Power-aware Source Routing (PSR) protocol for MANETs. This protocol was based on the Dynamic Source Routing protocol (RFC4728, n.d.). PSR computes the cost of routes while taking into consideration both transmission power and remaining battery power. In PSR, the source broadcasts a message and intermediate nodes compute the path cost and add it to the header of the broadcast message. The destination then adds the least cost path to the reply and sends it back to the source. This solution fits the needs of MANETs but because of its broadcast nature, it is not suitable for the more resource constrained sensor networks. Since most sensor network applications require static sensors and are more resource constrained than MANETs, the routing solutions that were developed for MANETs are not suitable for the low power sensor networks. As a result, the Routing Over Low power and Lossy networks (ROLL) group was created as part of IETF in 2008 (Watteyne & Richichi, 2010) to help develop a standardized routing solution for sensor networks.

In (Watteyne & Richichi, 2010), the authors define a set of criteria that routing protocols must possess for routing in low-power and lossy networks. These criteria consist of satisfactory performance in: 1. Routing state. 2. Loss response. 3. Control cost. 4. Link cost. 5. Node cost. The authors then conclude that none of the mature IETF protocols, that were developed for MANETs, fulfill those requirements. The protocols examined in this work are: OSPF (RFC2328, n.d.), IS-IS (RFC1142, n.d.), OLSR (RFC3626, n.d.), OLSRv2 (draft-ietf-manet-olsrv2-12, n.d.), TBRPF (RFC3684, n.d.), RIP (RFC2453, n.d.), AODV (RFC3561, n.d.), DYMO (draft-ietf-manet-dymo-mib-04, n.d.), DSR (RFC4728, n.d.), IPv6 Neighbor Discovery (RFC4861, n.d.) and MANET-NHDP (draft-ietf-manet-nhdp-15, n.d.). In (Watteyne & Richichi, 2010), the authors suggest that a new protocol specification document needs to be created for routing in low-power and lossy networks. The discussion in (Watteyne & Richichi, 2010) was limited to mature and well documented IETF protocols, in the remaining of this section, we examine "energy aware" routing protocols designed for wireless sensor networks that have not been included in this review.

Routing algorithms with energy considerations aim to either minimize the energy consumption of the networks as a whole or increase the lifetime of the network. Protocols that attempt to minimize the energy consumption of the network usually compute and use the shortest paths in the network. As a consequence, a few select motes are usually overused and their energy reserve is depleted earlier than the rest of the motes. This could result in the network becoming partitioned and could therefore end its useful lifetime prematurely.

Most applications of WSNs are deployed in remote areas and are scheduled to monitor the area for long periods of time. In this case, extending the useful lifetime of the network is of at most importance. The concept of 'lifetime of the network' is difficult to define in WSNs (Dietrick & Dressler, 2009). For practical purposes, we define the useful lifetime of the network as: 'The total amount of time that the network is able to do useful work'. If for example the purpose of the network is to record sensor readings from ten different areas for as long as possible, the useful (operational or functional) lifetime of the network will be the total amount of time that at least one sensor is functional in each of the ten different areas and that there exists a path between each of those senors to the sink, i.e., those sensor motes are able to relay their readings to the sink. The useful lifetime of the network is therefore application specific and a uniform definition may not apply to all types of WSN applications.

The shortcomings of the broadcast-based protocols have led to the design of data-centric routing mechanisms. One of the earliest works on this type of protocols is SPINS (Heinzelman et al., 1999) where the data is named using high-level descriptors (meta-data). In this case, sensors exchange meta-data. The protocol relies on three types of messages: 1. ADV message, which is used to advertise particular meta-data, 2. REQ message used to request specific data, and 2. DATA message used to deliver the actual data. Spins achieves significant energy savings over traditional broadcast-based protocols (a factor of 3.5) and reduces the data redundancy in half. However, Spins does not guarantee the delivery of data to the requesting node, which makes this protocol unpractical for several applications of WSNs (Akkaya & Younis, 2003).

In data-centric routing algorithms, regions of sensors are queried to send their sensed readings to the sink. Because of the redundancy in sensors in each region, the data needs to be aggregated before it is forwarded to the sink. Several algorithms have been proposed to perform data aggregation to disregard the redundant information. Sensor Protocols for

Information via Negotiation (SPIN) (Kulik, 1999) was the first work to suggest eliminating redundant information to save energy. Later, a series of protocols that use directed diffusion were proposed (Intanagonwiwat et al., 2000), (Braginsky & Estin, 2002), (Schurgers & Srivastava, 2001), (Chu et al., 2002).

An important step in routing in wireless sensor networks was the creation of routing algorithms based on directed diffusion, the first introduction is described in (Intanagonwiwat et al., 2000). In directed diffusion, a node sends a query for some particular data (data here is identified using an attribute-value pair). As a result, data matching the query description is "drawn" towards the querying node. The data can be aggregated by intermediate nodes and all the communication is only neighbor-to-neighbor. These types of algorithms achieve significant energy savings over the traditional broadcast-based algorithms. Despite the energy saving properties of the directed diffusion algorithms, they are not suitable for all sensor network applications. Some sensor network applications require continuous data flow from the sensors to the sink, as a consequence, query based algorithms will not be suitable for such applications since the sink would need to continuously query each sensor for data (Akkaya & Younis, 2003).

An alternate way of relaying information in WSNs, other than flooding, is gossiping (Kyasanur et al., 2006). In gossiping, the source node selects a random neighbor and forwards the data to them. The process continues until the destination is reached or a maximum number of hops is achieved. Similar to flooding protocols, gossiping protocols also waste energy by sending messages by sensors that cover overlapping areas. In addition, gossiping algorithms can suffer from excessive delays because the next hop node is selected randomly.

An improvement to the traditional gossiping protocols is the location-based protocols. In these protocols, location information is used to direct the routing in order to reduce the number of transmissions and therefore save energy. One such protocol is SPEED (He et al., 2002). This protocol uses a combination of feedback control and non-deterministic geographic forwarding to provide real-time unicast, area-multicast and real-time area-anycast.

In (Li et al., 2001), the authors propose an energy saving routing scheme called the adaptive max-minzPmin scheme. This routing algorithm selects a route that maximizes the minimum residual energy as long as it consumes no more than zPmin energy (Pmin energy is he amount of energy consumed by the minimum-energy route). This algorithm also computes the minimum node lifetime of the network and adjusts its routing criterion accordingly. While this method is hard to implement (keeping track of the lifetime of the nodes in a central location), it is more practical for ad hoc networks than it is for sensor networks.

Another family of protocols is the hierarchical routing protocols. The main purpose of creating a hierarchy within a sensor network is to achieve scalability, i.e., the network performance should decrease slowly in response to an increase in the network size. The main form of hierarchical routing in WSNs is clustering, which consists of organizing the nodes into clusters where each cluster has a cluster head. The cluster head is then in charge of performing data aggregation or forward the packets on to the next hop. This leads to a smaller amount of data being transmitted to the sink, which intrinsically saves energy.

One of the first clustering protocols, LEACH is described in (Heinzelman et al., 2000). LEACH randomly rotates the head cluster in order to balance the energy consumption amongst the nodes in the cluster and uses data fusion in order to reduce the amount of data sent to the sink.

As a result, LEACH achieves significant energy savings compared to conventional routing protocols. Several other hierarchical protocols have been proposed in the literature who build up on LEACH such as TL-LEACH (Loscri et al., 2005) which proposes a two-level hierarchy to LEACH, EECS (Ye et al., 2006) where nodes compete for the position of cluster head, HEED (Yonis & Fahmy, 2004) where cluster heads are selected based on the distance between nodes, among others.

In (Iwanicki & Steen, 2009), the authors develop a framework to test the various hierarchical routing protocols proposed for WSNs. The authors state that hierarchical routing is a promising solution for the resource constrained WSNs and caution that the theoretical results presented in most hierarchical work can be very different from the results obtained using a more realistic framework. The proposed framework dismisses the idea of rotating the cluster head to save energy because this change complicates route computation by changing the routing addresses. The authors conclude that there is no one optimal hierarchical routing protocol for all WSN applications, rather protocols are application and requirement dependent.

In conclusion, there still exists the need to develop a low-frills, low-power, manageable and adaptable protocol for routing in the resource constrained sensor networks. The ROLL working group is still working on a requirement specification document. They may in fact, not be able to propose a single protocol for all or most WSN applications and could end up proposing or extending more than one protocol.

4.4 Medium access control layer

The main duties of sensor motes are communication, sensing and computing. Amongst these three tasks, communication consumes the most energy. It is therefore imperative to make sure that the communication task is as efficient as possible in order to prolong the energy lifetime of the motes. It is the data link layer that is responsible for establishing communication links between the motes, allowing the motes to share the wireless medium fairly and detecting/correcting transmission errors. Power considerations at the data link layer involve studying the hardware of the communication module (see Section 3) , the implementation of protocols such as the power management protocol and manipulating the power level of the transceiver.

The most energy waste occurs when a mote receives multiple frames at the same time. In this case all the frames that collide need to be discarded which results in wasted transmissions and receptions and increased latency. Other causes of energy waste are control packet overhead, overhearing unnecessary traffic and the long idle time in WSNs. In fact, in WSNs, idle listening consumes more than half the amount of energy required for reception (Ye et al., 2004). The Medium Access Control (MAC) layer is the sublayer of the data link layer that is responsible for handling the contention over the medium (in this case, the wireless medium). The main media access protocols used in wireless networks are Time Division Multiple Access (TDMA), Frequency Division Multiple Access (FDMA), Carrier Sense Multiple Access (CSMA), Request To Send/Clear To Send (RTS/CTS) protocols, and the IEEE 802.11 protocol. The purpose of these schemes is to avoid channel contention. In the following, we review the most relevant MAC protocols that are proposed for use with wireless sensor networks. The channel contention scheme in these protocols is based on the above described contention prevention mechanisms. In the rest of this Section, we study the main MAC layer protocols that are proposed in the literature and analyze their power-saving properties.

In this work, we consider energy efficiency to be the most important attribute in a MAC protocol. Other important attributes for a MAC protocol consist of providing fair and efficient access to the medium, scalability and adaptability to change.

Most ad hoc network and WSN applications require the network to be deployed for an extended period of time. During their deployment, the motes are programmed to sense the environment and relay sensor readings to the sink. Several MAC protocols have been proposed for these applications where the motes are periodically scheduled to be in a power-saving state (a sleep state or an off state) in order to save their battery power and extend their deployment lifetime, see (Singh & Raghavendra, 1998; Stewart & Tragoudas, 2007; Ye et al., 2004) among others.

PAMAS (Singh & Raghavendra, 1998) is a MAC protocol based on RTS/CTS. PAMAS schedules sleep intervals for sensor nodes to avoid overhearing and uses separate channels for data and control frames. In PAMAS, nodes probe their neighbors for transmission time in order to avoid collision as well. PAMAS reduces energy consumption by avoiding collision and transmission overhearing at the expense of increased hardware complexity, which in turn affects the power consumption.

The S-MAC (Ye et al., 2004) protocol reduces the energy consumption of the nodes by implementing the following mechanisms. First, it reduces idle listening by scheduling sleep intervals for nodes, in fact, S-MAC coordinates sleep intervals amongst neighboring nodes. Second, it divides long messages into smaller packets and transmits them back to back. As a result, nodes with longer messages occupy the wireless medium for longer periods of time. The authors show that this seemingly "unfair" advantage results in energy savings over traditional "fair" methods. Third, it implements a low-duty-cycle that reduces idle listening. Finally, it uses in-channel signaling to reduce overhearing by extending the work from PAMAS (Singh & Raghavendra, 1998).

S-MAC's mechanisms do reduce energy consumption at the expense of increased message latency. However, the predefined sleep intervals limit the flexibility of the protocol and the broadcast mechanism increases the probability for collision because S-MAC does not use RTS/CTS (Demirkol et al., 2006).

TMAC (Van & Langendoen, 2003) is similar to SMAC except that each node is equipped with a timer. In TMAC, a node is put on the low-power/sleep state if it does not transmit or receive for the entire duration of the timeout period. TMAC performs significantly better than S-MAC under variable load.

In WiseMAC (El-Hoiydi & Decotignie, 2004), the authors propose a downlink (to be used when the sink transmits packets to sensors). WiseMAC uses non-persistent CSMA (np-CSMA) with preamble sampling in order to decrease idle listening. In this case, a preamble is used to alert the receiving node that a data packet is on its way. The preamble precedes each data packets. All the nodes in the medium listen to the medium for a constant time interval referred to as the sampling period. If a node hears a transmission while it is listening to the medium, it will continue to listen until it receives a frame or until the medium becomes idle. The sink precedes each data frame with a preamble sequence that is equal to the sampling period. This guarantees that the receiving node will be able to detect the transmission. On the downside, the long preamble sequence results in a low throughput and in increase power consumption. In addition, all the nodes within wireless range of the receiving node are able to hear the transmission. WiseMAC proposes an improvement to this where the sink takes advantage of knowing the sampling schedule of the nodes. The sink therefore, sends a smaller preamble and a frame right when the receiving node is scheduled to start sampling the medium.

WiseMAC suffers from two main drawbacks (Demirkol et al., 2006). The first drawback results from its decentralized sleep schedule where nodes wake up from their sleep cycle at different times. This is inefficient when broadcast communication is used because the broadcasted frames would need to be stored at the neighbors who are awake and end up being transmitted multiple times. The second drawback of the protocol is the fact that it is vulnerable to the hidden terminal problem where collision can happen at a node if it receives transmissions from two nodes that are not within transmission range of each other (Note that this is not a problem if WiseMAC is only used as a downlink protocol)

TRAMA (Rajendran et al., 2003) is a collision-free TDMA based MAC protocol for sensor networks. TRAMA ensures energy efficiency by avoiding collision during unicast, multicast and broadcast transmissions. In addition, in TRAMA, nodes can switch to a low-power state whenever they are not transmitting or receiving frames to save energy. In TRAMA, a node is elected to transmit within a two-hop neighborhood during each time slot. This mechanism avoids the hidden terminal problem.

TRAMA achieves significant energy savings due to: 1. the increased amount of low-power states, 2. the decreased amount of communication since the receiving nodes are indicated a priori, and 3. the significant decrease in collision probability. However, the latency when using TRAMA is longer compared to CSMA as a result of the high percentage of sleep time (Demirkol et al., 2006).

Berkeley MAC (B-MAC) (Polastre et al., 2004) is a low frills protocol based on clear channel assessment, it uses low power listening with preamble sampling. The default mode in B-MAC does not include a mechanism to avoid the hidden terminal problem, which could be implemented by higher layers if needed. B-MAC achieves significant energy savings when varying check time, by making the preamble constant and setting the sample rate. However, since the protocol is bare-bone, additional features would have to be implemented at higher layers when needed, which increases the complexity of the system as a whole.

Even though multiple MAC layer protocols provide adequate performance, no single protocol has been chosen as a standard. This is due to the fact that some protocols perform better than others for particular applications, communication pattern, network infrastructures and or network densities. An ideal energy efficient MAC layer protocol for WSNs would use a local schedule for motes to turn to the low-power/off state as a function of their residual energy as well as their sensing schedule. The schedule should aim to maximize the sleep time of the motes while preserving their sensing schedule, local connectivity and balancing their energy levels in order to increase the lifetime of the network as a whole.

4.4.1 Physical layer

Frequency detection, generation, modulation and coupling are the responsibility of the physical layer and are explained in detail in the hardware section. Note that when an engineer is charged with designing a physical layer, propagation effects due to the ambient conditions must be considered.

5. Conclusion and future work

This chapter reviews the hardware architecture of wireless sensor motes, as well as their protocol stack focusing on power considerations at every level. We conclude that because of the diversity in WSN applications, it is very difficult to derive a universal power efficient architecture both in terms of hardware and software.

As far as the hardware components in WSNs, many advances have been made over the last few years. These improvements include more efficient apertures with better directivity and lower VSWR. The sensor element has been made to become more resolute while power management has improved due to the accessibility of more exotic materials for energy storage. The future holds near perfect antenna with nearly a 1:1 VSWR ensuring most of the energy leaving the system goes where it's designed to propagate. Researchers at Purdue University are working toward ensuring optical sensors are near perfectly efficient with negative refractive metamaterials and photon collection efforts.

In terms of the WSN protocol stack, no one protocol has been adopted as a WSN standard, rather each protocol is designed to efficiently serve one or more WSN applications. The power efficiency of protocols has become the number one constraint in almost every layer of the protocol stack. More work is needed to design and develop protocols that are less application specific and still power efficient.

6. References

Kemal Akkaya and Mohamed Younis, *A Survey on Routing Protocols for Wireless Sensor Networks*. Elsevier journal of Ad Hoc Networks, Volume 3, pages 325-349.

I. F. Akyildiz, W. Su, Y. Sankarasubramaniam, and E. Cayirci, *Wireless sensor networks: a survey*. Elsevier Computer Networks, Volume 38, pages 393-422, 2002.

ALERT, www.alertsystems.org

S. Basagni, M. Conti, S. Giordano, Iv. Stojmenovic, *Chapter 11: Energy-efficient Communication in ad hoc Wireless Networks*. Mobile ad hoc networking, Wiley-IEEE Press, 2004.

Bontempi, G., and Le Borgne, Y. (2005). *An adaptive modular approach to the mining of sensor network data*. In Workshop on Data Mining in Sensor Networks, SIAM SDM, Newport Beach, CA, USA, April.

D. Braginsky and D. Estin, *Rumor routing algorithm for sensor networks*. Proceedings of the first workshop on Sensor Networks and Applications (WSNA), Atlanta, GA, October 2002.

Wenshan Cai, and Vladmir Shalaev, *Optical Metamatrials: Fundamentals and Applications*. Springer, 2010.

M. Chu, H. Haussecker, and F. Zhao, *Scalable Information-Driven Sensor Querying and Routing for ad hoc Heterogeneous Sensor Networks*. The International Journal of High Performance Computing Applications, Vol. 16, No. 3, August 2002.

C. Coue, T. Franichard, P. Bessiere, and E. Mazer, *Multi-sensor data fusion using Bayesian programming: An automotive application*. Proceedings of the IEEE/RSJ International Coference on Intelligent Robots and Systems. Vol. 1, Lausanne, Switzerland, pages 141-146.

Crossbow Technologies, *http://www.xbow.com/*

Ilker Demirkol, Cem Ersoy, and Fatih Alagoz, *MAC Protocol for Wireless Sensor Networks: a Survey*. IEEE Communications Magazine, 2006.

A. P. Dempster, *A generalization of Bayesian inference*. J. Royal Stat. Soc., Series B 30, pages 205-247, 1968.

Isabel Dietrich and Falko Dressler, *On the lifetime of wireless sensor networks*. ACM Transactions on Sensor Networks, Vol. 5, No. 1, Article 5, February 2009.

Draft-IETF-MANET-DYMO-mib-04: http://tools.ietf.org/html/draft-ietf-manet-dymo-mib-04

Draft-IETF-MANET-NHDP-15: http://tools.ietf.org/html/draft-ietf-manet-nhdp-15

Draft-IETF-MANET-olsrv2-12: https://datatracker.ietf.org/doc/draft-ietf-manet-olsrv2/

A. El-Hoiydi and J.-D. Decotignie, *WiseMAC: An Ultra Low Power Protocol for the Downlink of Infrastructure Wireless Sensor Networks*. Proceedings of the Ninth IEEE Symposium on Computers and Communication, ISCCÕ04, pages 244-251, Alexandria, Egypt, June 2004.

E. H. Elhafsi, N. MItton, D. Simplot-Ryl, *End-to-End Energy Efficient Geographic Path Discovery With Guaranteed Delivery in Ad Hoc and Sensor Networks*. 19th Annual International Symposium on Personal, Indoor and Mobile Radio Communications (PIMRC). Cannes, France: IEEE, 15-18 Septembr 2008, pp. 1-5.

Jeremy Elson, Lewis Girod, and Deborah Estrin, *Fine-grained network time synchronization using reference broadcasts*. Proceedings of the ACM OSDI'02, Boston, MA, December 2002.

Mehdi Esnaashari and M.R. Meybodi. *Data Aggregation in Sensor Networks using Learning Automata*. Wireless Networks 20

S. Ganeriwal, R. Kumar, and M Srivastava, *Timing Sync Protocol for Sensor Networks*. Proceedings of the ACM SenSys'03, 2003.

J.V. Greunen, and J. Rabaey, *Lightweight Time Synchronization for Sensor Networks*. Proceedings of the 2nd ACM International Conference on Wireless Sensor Networks and Applications (WSNA'03), San Diego, CA, September 2003.

S. Grime, H.F. Durrant-Whyte, *Data fusion in decentralized sensor networks*. Control Eng. Practices, 2(5):849-63, 1994.

N. Guilar, A. Chen, T. Kleeburg, and R. Amirtharajah, *Integrated Solar Energy Harvesting and Storage*. Proceedings of the International Symposium of Low Power Electronics and Design (ISLPED'06), October 2006.

I. Gupta, D. Riordan, and S. Sampalli, *Cluster-head election using fuzzy logic for wireless sensor networks*. Proceedings of the 3rd Annual Communcation Networks and Services Research Conference (CNSR'05). IEEE, Halifax Canada, pages 255-260.

Tian He, John Stankovic, Chenyang Lu, and Tarek Abdelzaher, *SPEED: A Real-Time Routing Protocol for Sensor Networks*.

Heinzelman, W., Chandrakasan, A., and Balakrishnan, H. (2000). *Energy-efficient communication protocol for wireless microsensor networks*. In Proceedings of 33rd Hawaii International Conference on System Science (HICSS Õ00), January.

W.B. Heinzelman, A. P. Chandrakasan, and H. Blakrishnan, *An Application-Specific Protocol Architecture for Wireless Microsensor Neworks*. IEEE Transactions on Wireless Communications, vol. 1, no. 4, pp. 660-670, October 2002.

W. Heinzelman, J. Kulik, and H. Balakrishnan, *Adaptive protocols for information dissemination in wireless sensor networks*. Proceedings of the 5th annual ACM/IEEE International Conference on Mobile Computing and Networking (MobiComÕ99), Seattle, WA, August 1999.

W. Heinzelman, A. Chandrakasan, and H. Balakrishnan, *Energy-Efficient communication protocol for wireless sensor networks*. Proceedings of the Hawaii International Conference System Sciences, Hawaii, January 2000.

Fei Hu, and Xiaojun Cao, *Wireless Sensor Networks: Principles and Practice*. CRC Press, 2010.

C. Intanagonwiwat, R. Govindan and D. Estin, *Directed Diffusion: A scalable and robust communication pradigm for sensor networks*. Proceedings of the 6th annual ACM/IEEE International Conference on Mobile Computing and Networking (MobiComÕ00), Boston, MA, August 2000.

Sandy Irani, Sandeep Shukla, and Rajesh Gupta, *Power Savings* . Proceedings of ACM Transactions on Algorithms, Vol. 3, No. 4, Article 41, November 2007.

Konard Iwanicki, and Maarten van Steen, *On Hierarchical Routing in Wireless Sensor Networks*. Proceedings of IPSN'09, April 13-16, 2009, San Francisco, California, USA.

R. E. Kalman, *A new approach to linear filtering and prediction problems*. Transactions. ASME J. Basic Engin. 82, pages 35-45, 1960.

Mauritus Morne, *Reliable Ad-hoc On-demand Distance Vector Routing Protocol*. Proceedings of the International Conference on Networking, International Conference on Systems and International Conference on Mobile Communications and Learning Technologies (ICNICONSMCL'06).

Joanna Kulik, Wendi Rabiner, Hari Balakrishnan, *Adaptive Protocols for Information Dissemination in Wireless Sensor Networks*. Proceedings of the 5th ACM/IEEE Mobicom Conference, Seattle, WA, August 1999.

P. Kyasanur, R. R. Choudhury, and I. Gupta. *Smart Gossip: An Adaptive Gossip-based Broadcasting Service for Sensor Networks*. Proceedings of MASS'06, 2006.

Andreas Lachenmann, Pedro Jos'e Marr'on, Matthias Gauger, Daniel Minder, Olga Saukh, and Kurt Rothermel, *Removing the Memory Limitations of Sensor Networks with Flash-Based Virtual Memory*. Proceedings of the 2nd ACM SIGOPS/EuroSys European Conference on Computer Systems 2007.

Xu, Y., Lee, W. C., Xu, J., and Mitchell, G. (2006). *Processing window queries in wireless sensor networks*. In IEEE International Conference on Data Engineering (ICDEÕ06), Atlanta, GA, April.

Christoph Lenzen, Philipp Sommer, and Roger Wattenhofer, *Optimal clock synchronization in network*. Proceedings of the the 7th ACM Conference on Embedded Networked Sensor Systems (SenSys'09).

Christoph Lenzen, Philipp Sommer and Roget Wattenhofer, *Optimal Clock Synchronization in Networks*. Proceedings of SensSysÕ09, November 4-9, 2009, Berkeley, CA, USA.

P. Levis, A. Tavakoli and S. Dawson-Haggerty, *Overview of Existing Routing Protocols for Low Power and Lossy Networks*. IETF ROLL, IETF draft, 14 February 2009, Draft-ietf-roll-protocols-survey-07.

Qun Li, Javed Aslam, and Daniela Rus, *Online power-aware routing in wireless ad-hoc networks*. Proceedings of the 7th Annual International Conference on Mobile Computing and Networking, 2001.

Liqun Li, Guoliang Xing, Limin Sun, Wei Huangfu, Ruogu Zhou, and Hongsong Zhu, *Exploiting FM Radio Data System for Adaptive Clock Calibration in Sensor Networks*. Proceedings of the ACM MobiSys'11, Washington DC, June 28, 2011.

Liu, C., Wu, K., and Pei, J. (2005). *A dynamic clustering and scheduling approach to energy saving in data collection from wireless sensor networks*. In Proceedings of the Second Annual

IEEE Communications Society Conference on Sensor and Ad Hoc Communications and Networks (SECONÕ05), Santa Clara, California, USA, September.

V. Loscri, G. Morabito, and S. Marano, *A Two-Level Hierarchy for Low-Energy Adaptive Clustering Hierarchy*. Proceedings of the Vehicular Technology Conference (VTC'05), September 25-28, 2005.

Lotfinezhad, M., and Liang, B. (2004). *Effect of partially correlated data on clustering in wireless sensor networks*. In Proceedings of the IEEE International Conference on Sensor and Ad hoc Communications and Networks (SECON), Santa Clara, California, October.

Saoucene Mahfoudh, and Pascale Minet. *Energy-aware Routing in Wireless Ad Hoc and Sensor Networks*, Proceedings of the 6th International Wireless Communications and Mobile Computing Conference (IWCMCÕ10). June 2010.

Alan Mainwaring, Joseph Polastre, Robert Szewczyk, David Culler, and John Anderson. *Wireless Sensor Networks for Habitat Monitoring*. Proceedings of WSNA'02, September 28, 2002, Atlanta, Georgia.

M. Marina, and S. Das, *On-demand Multipath Distance Vector Routing in Ad Hoc Networks*. Proceedings of the 2001 IEEE International Conference on Network Protocols (ICNP), pages 14-23, IEEE Computer Society Press, 2001.

M. Maroti, B. Kusy, G. Simon, and A. Ledeczi, *The Flooding Time Synchronization Protocol*. Proceedings of the 2nd ACN Conference on Embedded Networked Sensor Systems (SenSys'04), Baltimore, Maryland, 2004, pages: 39-49.

Morteza Maleki, Karthik Dantu, and Massoud Pedram *Power-aware Source Routing Protocol for Mobile Ad Hoc Networks*. Proceedings of the ISLPED'02, August 12-14, 2002, Monterey, California, USA.

Eduardo Nakamura, and Alejandro Frery. *Information Fusion for Wireless Sensor Networks: Methods, Models, and Classifications*, Computing Surveys (CSUR), Volume 39, Issue 3, 2007.

Evgenii E. Narimanov, and Alexamder V. Kildishev, *Optical black hole: broadband omnidirectional light absorber*. Appl. Phys. Lett. 95, 041106 (2009).

NTP: http://www.ntp.org/

OLSR: RFC 3626

D. Petriovic, R. C. Shah, L. Ramchandran, and J. Rabaey, *Data funneling: Routing with aggregation and compression for wireless sensor networks*. Proceedings of the first IEEE International Workshop on Sensor Network Protocols and Applications (SNPA'03). IEEE, Anchrage, AK, pages 156-162.

Kris Pister, *Autonomous sensing and communication in a cubic millimeter*. Http://www-bsac.eecs.berkeley.edu/ pister/SmartDust/

Joseph Polastre, Jason Hill, and David Culler, *Versatile low power media access for wireless sensor networks*. Proceedings of SenSys'04, 2004.

Octavian Postolache, Pedro Silva Girao, and Jose Miguel Dias Pereira, *Non-Volatile Memory Interface Protocols for Smart Sensor Networks and Mobile Devices*. Data Storage, InTech publishers, April 2010.

V. Rajendran, K. Obraczka, J. J. Garcia-Luna-Aceves, *Energy-Efficient, Collision-Free Medium Access Control for Wireless Sensor Networks*. Proceedings of SenSysÕ03, November 5-7, 2003, Los Angeles, California, USA.

RFC 1142: http://tools.ietf.org/html/rfc1142

RFC 2328:http://www.ietf.org/rfc/rfc2328.txt

RFC 2453: http://tools.ietf.org/html/rfc2453

RFC 3561: http://www.ietf.org/rfc/rfc3561.txt

RFC 3626:http://www.ietf.org/rfc/rfc3626.txt

RFC 3684: http://www.ietf.org/rfc/rfc3684.txt

RFC 4728: http://www.ietf.org/rfc/rfc4728.txt

RFC 4861: http://tools.ietf.org/html/rfc4861

Kay Romer, *Time synchronization in ad hoc networks*. Proceedings of the ACM MobiHoc'01, Long Beach, CA October 2001.

Rosemark, R., and Lee, W. C. (2005). *Decentralizing query processing in sensor networks*. In The Second International Conference on Mobile and Ubiquitous Systems: Networking and Services (MobiquitousÕ05), San Diego, CA, July (pp. 270Ð280).

C. Schurgers and M.B. Srivastava, *Energy efficient routing in wireless sensor networks*. MILCOM Proceedings on Communications for Network-Centric Operations: Creating the Information Force, McLean, VA 2001.

G. Shafer, *A Mathematical Theory of Evidence*. Princelton University Press, Princeton, NJ 1976.

Vladimir M. Shalaev, Wenshan Cai, Uday K. Chettiar, Hsiao-Kuan Yuan, Andrey K. Sarychev, Vladimir P. Drachev, and Alexander V. Kildishev, *Negative index of refraction in optical metamaterials*. Optics Letters, Vol. 30, Issue 24, pages 3356-3358, 2005.

S. Singh and C.S. Raghavendra, *Power aware multi-access protocol with signaling for ad hoc networks*. ACM Computer Communication Review Vol. 28 No. 3 (July 1998) pages. 5-26

Philipp Sommer and, Roger Wattenhofer, *Gradient Clock Synchronization in Wireless Sensor Networks*. Proceedings of the International Conference on Information Processing in Sensor Networks, 2009.

Soro, S., and Heinzelman, W. (2005). *Prolonging the lifetime of wireless sensor networks via unequal clustering*. In Proceedings of the 5th International Workshop on Algorithms for Wireless, Mobile, Ad Hoc and Sensor Networks (IEEE WMAN Õ05), April.

K. Stewart, and S. Tragoudas, *Managing the power resources of sensor networks with performance considerations*. Computer Communications Journal, Volume 30, Number 5, pages:1122-1135, March 2007.

A. Tsymbal, S. Puuronen, and D. W. Patterson, *Ensemble feature selection with the simple Bayesian classification*. Information Fusion 4, 2, June, pages 87-100.

T. van Dam, and K. Langendoen, *An adaptive energy-efficient mac protocol for wireless sensor networks*. Proceedings of the First ACM Conference on Embedded Networked Sensor Systems, November 2003.

Natalia Vassileva, Francisco Barcelo-Arroyo, *A Survey of Routing Protocols for Maximizing the Lifetime of Ad Hoc Wireless Networks*. International Journal of Software Engineering and its Applications, Vol. 2, No. 3, July 2008.

Virrankoski, R., and Savvides, A. (2005). *TASC: Topology adaptive spatial clustering for sensor networks*. In Second IEEE International Conference on Mobile Ad Hoc and Sensor systems, Washington, DC, November.

Thomas Watteyne, Maria Grazia Richichi, *From MANET to IETF ROLL Standardization: A Paradigm Shift in WSN Routing Protocols*, submitted to IEEE Communications Surveys and Tutorials.

Jon Wilson, *Sensor Technology Handbook*. Elsevier, ISBN: 0-7506-7729-5, December 2004.

Winter, J., Xu, Y., and Lee, W. C. (2005). *Energy efficient processing of K nearest neighbor queries in location-aware sensor networks*. In The Second International Conference on Mobile and

Ubiquitous Systems: Networking and Services (MobiquitousÕ05), San Diego, CA, July (pp. 281Ð292).

L. Xiao, S. Boyd, and S. Lall, *A scheme for robust distributed sensor fusion based on average consensus*. Proceedings of the 4th International Symposium on Information Processin in Sensor Networks (IPSN'05), pages 63-70, 2005.

Z. Xiong, A. D Liveris, and S. Cheng, *Distributed source coding for sensor networks*. Proceedings of IEEE Sig. Proc. Mag. 21, 5, September 2004, pages 80-94.

M. Ye, C. Li, G. Chen, and J. Wu, EECS: *An Energy Efficient Clustering Scheme in Wireless Sensor Networks*. Ad Hoc and Sensor Wireless Networks, Vol. 3, Pages 99-119, April 2006.

W. Ye, J. Heidemann, D. Estrin, *Medium Access Control With Coordinated Adaptive Sleeping for Wireless Sensor Networks*. IEEE/ACM Transactions on Networking, Volume 12, Issue: 3, Pages: 493-506, June 2004.

Zhenzhen Ye, Alhussein Abouzed, and Jing Ai. *Optimal Stochastic Policies for Distributed Data Aggregation in Wireless Sensor Networks*, IEEE/ACM Transactions on Networking, VOL. 17, NO. 5, October 2009.

O. Younis and S. Fahmy, *HEED: A Hybrid Energy-Efficient Distributed Clustering Approach for Ad Hoc Sensor Networks* IEEE Transactions on Mobile Computing, vol. 3, no. 4, Oct-Dec 2004.

Younis, O., and Fahmy, S. (2005). *An experimental study of routing and data aggregation in sensor networks*. In Proceedings of theInternational Workshop on Localized Communication and Topology Protocols for Ad hoc Networks (LOCAN), held in conjunction with The 2nd IEEE International Conference on Mobile Ad Hoc and Sensor Systems (MASS-2005), November.

8

Review of Optimization Problems in Wireless Sensor Networks

Ada Gogu[1], Dritan Nace[1], Arta Dilo[2] and Nirvana Meratnia[2]

[1]*Université de Technologie de Compiègne*
[2]*University of Twente*
[1]*France*
[2]*The Netherlands*

1. Introduction

Wireless Sensor Networks (WSNs) are an interesting field of research because of their numerous applications and the possibility of integrating them into more complex network systems. The difficulties encountered in WSN design usually relate either to their stringent constraints, which include energy, bandwidth, memory and computational capabilities, or to the requirements of the particular application. As WSN design problems become more and more challenging, advances in the areas of Operations Research (OR) and Optimization are becoming increasingly useful in addressing them.

This study is concerned with topics relating to network design (including coverage, topology and power control, the medium access mechanism and the duty cycle) and to routing in WSN. The optimization problems encountered in these areas are affected simultaneously by different parameters pertaining to the physical, Medium Access Control (MAC), routing and application layers of the protocol stack. The goal of this study is to identify a number of different network problems, and for each of these network problems to examine the underlying optimization problem. In each case we begin by presenting the basic version of the network problem and extend it by introducing new constraints. These constraints result mainly from technological advances and from additional requirements present in WSN applications. For all the network problems discussed here a wide range of algorithms and protocols are to be found in the literature. We cite only some of these, since we are concerned more with the network optimization problem itself, together with its different versions, than with a state of art of methods for solving it. Moreover, the cited methods have originated in a variety of disciplines, with approaches ranging from the deterministic to the opportunistic, including computational geometry, linear, nonlinear and dynamic programming, metaheuristics and heuristics, game theory, and so on. We go on to discuss the complexity inherent in different optimization problems, in order to give some hints to WSN designers facing new but similar scenarios. We try to highlight distributed solutions and information that is required to implement these schemes. For each topic the general presentation scheme is as follows:

i) Present the network problem

ii) Identify the relevant optimization problem

iii) Discuss the theoretical complexity of the optimization problem

iv) Describe some representative solution methods, including distributed methods

The relations between the two areas of WSN network design and OR have been discussed in some other works (Li, 2008; Nieberg, 2006; Ren et al., 2006; Suomela, 2009). In (Li, 2008; Nieberg, 2006; Suomela, 2009) the goal is to relate a network problem to its corresponding optimization problems and to discuss related questions in the OR literature that might feature in a solution. For example, Suomela (2009) is focused on data gathering and scheduling problems in WSN. He identifies the respective optimization problems and presents some nice properties that a graph should have (e.g. bipartite, graph with unique identifiers, planar, spanners, etc) to facilitate the design of distributed algorithms for these optimizations problems. Ren et al. (2006) present a survey highlighting certain methodologies from operational research and the corresponding network problems that they can solve. In particular they relate *graph theory and network flow problems* to routing problems in WSN, *fuzzy logic* to clustering, and *game theory* to the problem of bandwidth allocation. Following on from these works we attempt to enlarge the spectrum of the network problems addressed, and for each network problem we highlight the optimization problem together with some effective methods proposed in the literature. Furthermore, we report at the end the study a discussion on open issues.

This chapter is organized as follows. The second section introduces certain methods from OR which are used to solve problems in WSN. The goal is to familiarize the reader with both the terminology and methods that are encountered in the OR domain and we refer to in the reminder of the study. In the third section we discuss several problems of WSN design, most of which must be addressed in the setup phase of the network. The fourth section is concerned with the routing problems. We report a classification of most used models and focus on how each of them is useful in addressing routing problems in WSN. The final section identifies some open issues in WSN and gives concluding remarks.

2. Operations research methodology used in WSN design

This section aims to introduce the reader to OR terminology and some representative solution methods from OR that are already used in WSN design. An Optimization Problem (OP) in OR is composed of two main parts. One is the objective/cost function to be maximized/minimized, and the second is concerned with the associated constraints that determine the feasibility domain. A solution of the OP is feasible if it satisfies all the constraints. From computational complexity point of view an OP is said to be polynomial if there exists a polynomial-time algorithm for solving it, otherwise it falls into NP-hard problems class. The solution methods used to solve the OP can be classed into two groups: exact methods and heuristic methods.

1. **Exact methods** seek a global optimal solution (if it exists) for the problem. The most familiar techniques among the exact methods commonly used for OPs in WSN are Linear, Nonlinear and Dynamic Programming. A general linear programming (LP) formulation is as follows:

$$\max \, cx \tag{1}$$
$$Ax \leq b \tag{2}$$
$$x \geq 0 \tag{3}$$

where A is a matrix, b and c are vectors giving respectively the right-hand terms and the cost coefficients, and x is the decision variable vector.

In cases where some decision variables have integer values while others have continuous values we refer to the problem as *Mixed Integer Linear Programming*. If, on the other hand, the vector x contains only integer values, then we have a case of *Integer Linear Programming (ILP)*. Note that the difficulty of the problem increases when we are dealing with ILP rather than LP, since ILP problems are commonly NP-hard. The most frequently used algorithms for solving LP problems are Simplex and Interior Points methods (Dantzig, 1963; Karmarkar, 1984), whereas for ILP problems there are Branch-and-Bound, Branch-and-Cut and Cutting Planes methods. Besides maximizing/minimizing an objective function, LP can be adapted so that it also guarantees fairness. In this case the objective function becomes a *max-min (or min-max)* objective function. In WSN we may often encounter network problems modeled according to this structure. It also happens that in some networks modeled by LP the number of variables is infinite or finite but huge, making an explicit enumeration impossible. In these cases the problem is solved iteratively. At each iteration new variables that potentially would lead to better solutions are generated by a method called column generation. The problem is solved when no new variables can be generated. Finally, when the objective function, or at least one of the constraints, is a nonlinear function, the problem becomes a nonlinear programming problem. In this type of problem the nature of the objective function is very important. If it is a convex function, then the problem is a nonlinear convex programming problem, where the best-known techniques include subgradient and Lagrangian decomposition (Kuhn, 1951; Shor, 1985). The above linear programming problems can be solved using a commercial solver such as CPLEX, Xpress-MP, etc. For nonlinear nonconvex programming the optimization becomes difficult and the solution methods less tractable. Another method worth citing is *Dynamic programming* (Bellman, 1957). This is a sequential approach where the decisions are taken optimally, step-by-step, until the complete solution has been constructed. This method works for problems that can be divided into subproblems that are simpler to solve and whose solutions will produce the global solution.

2. **Heuristic methods** are an important class of solution methods for practical optimization problems in WSN exhibiting high computational complexity. These approaches are intended to quickly provide near-optimal solutions to difficult optimization problems that cannot be solved exactly. Their advantages include easy implementation, rapidly-obtained solutions and robustness to variations in problem characteristics. However, in most cases they cannot guarantee the quality of the solution produced. Heuristic methods include local improvement methods that perform searches within the neighborhood of a feasible solution to the problem, and improve/construct the solution step by step by taking the best local optimal decision at each step. The main danger here is getting trapped at a local optimum, and to overcome this danger these methods may be combined with random approaches, multi-start approaches, and so on.

Similarly, metaheuristics are very general approaches used to guide other methods or procedures towards achieving reasonable solutions. Metaheuristics aim at reducing the search space and avoiding local optima. Most metaheuristics are life-inspired approaches such as Tabu Search (Glover, 1989), Evolutionary/Genetic algorithms (EA/GA) (Holland, 1975), Memetic algorithms (Moscato, 1999), Ant Colony Optimization (ACO) (Dorigo et al., 1996), and Particle Swarm Optimization (PSO) (Kennedy & Eberhart, 1995). Tabu Search starts with one feasible solution and constructs its neighborhood out of members that are obtained by permuting the elements of the feasible solution. The objective function is next calculated for each member of the neighborhood and the best one is selected. The process is then repeated but with the newly selected member as the starting point. An important element in this algorithm is loop-avoidance, meaning that it must not return to a solution that has been already processed, and for this reason all the forbidden movements are saved in a tabu list. In evolutionary or genetic algorithms the solutions of the problem are called individuals. A relatively small set of individuals selected within the enormous search space of the optimization problem are chosen to form the population. The population evolves during the iterations in a certain order known as generations. Genetic operators such as mutation and crossover are applied to produce better individuals. Their performance is evaluated based on a fitness/cost function. The algorithm stops when the solution is close to the optimum, or when a specific number of generations has been reached. Memetic algorithms combine GA with a local search. These algorithm follow the logic of a GA, but before applying genetic operators, every individual carries out a local search with the aim of improving its fitness. In ant colony optimization, an ant starts from a random node in the graph and selects the next node based on Equation (4).

$$P_{ij} = \begin{cases} \dfrac{\tau_{ij}^{\alpha} \cdot \eta_{ij}^{\beta}}{\sum_{k \in Liste} \tau_{ik}^{\alpha} \cdot \eta_{ik}^{\beta}} & \text{if } j \in List, \\ 0 & \text{otherwise} \end{cases} \tag{4}$$

where P_{ij} is the probability of choosing node j when the current node is i, τ_{ij} is the pheromone value of edge (i, j), η_{ij} the heuristic value, List contains all possible nodes accessible by the ant, and α, β are constants whose values depend in some way on the nature of the problem. In order to use an ACO algorithm for an OP, it is really important to present meaningfully the pheromone and heuristic values. When an ant passes through a node/edge, it deposits a pheromone value τ_{ij} in it. This value has to be proportional to the quality of the solution and it will help to attract other ants from the colony. The intention is that all the ants end up following the same trail, which hopefully represents the optimal solution. In order to avoid local optima this algorithm contains a process known as evaporation which periodically reduces the pheromone value deposited on a trail. The PSO algorithm imitates the flocking of birds. It initializes a number of agents (birds) and attaches two parameters, position and velocity (the velocity is given by two vectors which have orthogonal directions), to each of them. At each iteration the algorithm has to evaluate the positions of the agents and determine the subsequent positions, while accelerating their movement toward "better" solutions.

3. Network design issues

WSN design has to address a number of challenging factors. These include node deployment and coverage, connectivity and fault tolerance. The overall aim is always to lower costs, reduce the power consumption of the wireless environment and ensure a reliable network. Node deployment is the first essential stage in WSN design, and it strongly impacts the performance of the network as regards accurate event detection and efficient communication. Once the node is deployed, the problems of network organization become crucial, with topology and power control problems on one side, and medium access and scheduling strategies on the other. Solving these problems is an integral part of the design of a viable, energy-efficient network.

3.1 Optimal sensor deployment and coverage

WSN applications have particular requirements to satisfy, and one in common for all of them is coverage. The problem of maximizing the coverage of a given monitoring area has received a lot of attention in the literature. In this subsection we focus on three main problems related to this topic. First we discuss the problem of the minimum number of sensors required to cover a given area and guarantee network connectivity. The second problem is finding the best locations for a finite number of sensor nodes when seeking to satisfy the requirement of event detection. The third problem is identifying the regions that are not covered by sensors, assuming that the deployment is known.

The WSN deployment (or layout) problem is concerned with minimizing the number of deployed sensor nodes while ensuring the full coverage and connectivity of the monitoring area. As presented in (Efrat et al., 2005), the problem is a version of the Art Gallery problem, which is known to be NP-hard. The Art Gallery problem involves placing the smallest number of guards in an area such that every point in it can be surveyed by the guards. In this work Efrat et al. (2005) also show that the problem of deciding whether k sensors are sufficient to survey a region such that every point within the region is covered by three sensors is NP-hard. They propose an approximation algorithm based on geometry calculations for solving the problem.

However, most of the algorithms proposed for the layout problem derive from metaheuristic and heuristic methods. The work of Rotar et al. (2009) uses a new algorithm known as the Guided Hyper-plane Evolutionary Algorithm (GHEA). GHEA behaves basically the same as a multi-objective evolutionary algorithm manipulating a population and individuals. Whereas in the evolutionary algorithms the individuals are evaluated according to a fitness function, the novelty of GHEA lies in its evaluation of the population based on the hyperplane. The hyperplane will consist of points in the space which have better performances than any individual within the current population. Fidanova et al. (2010) propose an ant colony algorithm for this problem. As previously mentioned, ACO algorithms emulate the behavior of real ant colony where the greater the number of ants following a trail, the more attractive the trail becomes. In this case the area is modeled as a grid and all the points on the grid (or nodes) represent the search space. In order to apply the ant algorithm for the layout problem, from Equation (4) it is necessary to calculate the pheromone and the heuristic value every time that an ant passes through a node. The heuristic value attempts to reflect the best candidate node for the future movement of the ant (the new sensor placement) based on local information such as the number of grid points that the new candidate covers, whether the new candidate is reachable at a given distance, which is determined by the sensor transmission range, and finally whether this new position has already been selected by another ant. The pheromone, on the other hand, is initialized with a small value (e.g. the inverse of the number of ants) and for the upcoming iterations its value is updated according to the best solution value of the previous iteration.

In terms of the quality of service, attempts are made to find areas of lower observability from sensor nodes and to detect breach regions. The problem known as the Sensor Location Problem (SLP), formulated by (Cavalier et al., 2007), can be stated as follows: given a planar region, a given number of sensor nodes need to be positioned so that the probability of detecting an event in this region is maximized. The non-detection probability is expressed as a function of the distance between the sensor and a given position in the space where an event may take place, while the objective function aims to minimize the maximum of this product. In this formulation the problem is a difficult nonlinear nonconvex programming problem. Cavalier et al. (2007) proposes a heuristic algorithm that uses Voronoi polygons to estimate the probability of non-detection and to determine a search direction. The heuristic begins with an initial solution of m sensor locations $(x_1, x_2, ...x_m)$, on the basis of which the Voronoi diagram is constructed (see Fig. 1(a) and (b)). The construction of the Voronoi diagram must also take into account the area of the region. For every node the algorithm determines the point in the Voronoi polygon with the highest probability that an event will not be detected, and defines these points as the new node locations. The process is repeated until no further improvement is possible. We note that a similar problem is encountered by the wireless communication community in GSM networks and content-distribution wired networks (CDN). In GSM networks the problem is to find an optimal deployment of base stations within a region so that it provides maximum possible coverage. In CDN the problem is to determine the locations of proxies where the popular streams can be cached. This problem turns out to be the classical weighted $p - center$ location problem, where the objective is to locate p identical facilities that minimize the maximum weighted distance between clients and their corresponding (closest) facilities, assuming that each client is served by the closest facility (Averbakh & Berman, 1997). The $p - center$ problem is slightly different from SLP (note that for the SLP problem the clients correspond to events and the facilities correspond to sensors). A $p - center$ solution gives an assignment because each demand is assigned to a facility, while in SLP the event point (demand) can be visible to more than one sensor node (facility).

Once the sensors are deployed, coverage describes how well the sensors observe their target area or certain moving targets within this target area. In this context we need to know the path, known as the maximal breach path, that minimizes the maximum distance between every point on the path and its nearest sensor node. In other words this path represents the shortest path connecting the two endpoints which remains as far away as possible from sensor nodes. It was shown in (Duttagupta et al., 2008) that this problem is NP-hard. Most works in the literature propose methods relying on computational geometry and graph theory. Meguerdichian et al. (2001) suggest constructing the Voronoi diagram for the set of nodes in order to compute the maximal breach path. The edges of the Voronoi diagram provide the points of space which are at the greatest distance from the given set of sensors. These edges are weighted according to their distance from the nearest sensor. In this graph, the maximal breach path is a path maximizing the weight of its edges. A breadth-first-search (BFS) algorithm is then applied to find the maximal breach path. The Voronoi diagram and the maximal breach path are depicted in Fig. 1.

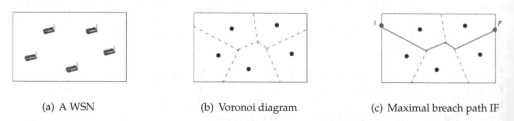

(a) A WSN   (b) Voronoi diagram   (c) Maximal breach path IF

Fig. 1. Voronoi diagram (b) for WSN nodes shown in (a), and maximal breach path (c).

3.2 Topology control

Node deployment can give rise to dense networks where sensors can have multiple potential neighboring nodes in common. This situation may lead to congestion and energy waste. To overcome this problem, topology control techniques are used to reduce the initial topology by choosing a subset of nodes having some property. Here the problem is finding a strongly connected subset of nodes that covers the rest of the nodes, so as to guarantee the connectivity of the whole network. This subset will be the backbone of the network, and every node excluded from it must have at least one edge in common with a node belonging to the subset. There are a number of advantages in obtaining a backbone topology, since for instance it may i) reduce network traffic by performing data aggregation and in-network processing, ii) avoid packet collisions as only the backbone nodes will forward packets to the sink while improving network throughput, and iii) make it possible to turn off the non-backbone nodes to save energy. This subsection will discuss the optimization problem for constructing the reduced topology, and the special case in which the lossy links are taken into account.

The problem is modeled as a widely-known mathematical problem called the Connected Dominating Set (CDS). A Dominating Set of a graph $G(V_{nodes}, E_{edges})$ is the subset of nodes $D \subset V$, such that every node that does not belong to D has at least one link in common with a node in D. In the special case in which these nodes have to be connected, the set is called the Connected Dominating Set (CDS). For many applications the smallest dominating set is sought, which brings us to the problem of finding the Minimum Connected Dominating Set (MCDS). The nodes in a CDS are called dominators, while other nodes are called dominatees. The MCDS problem is known to be NP-hard and is of the same difficulty and directly convertible to the vertex cover problem, the independent set computation problem, or the maximum clique problem. Yuanyuan et al. (2006) propose a two-phase method for obtaining

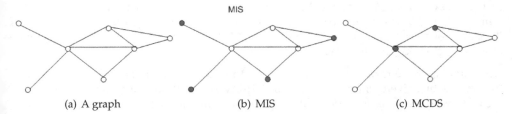

(a) A graph (b) MIS (c) MCDS

Fig. 2. Construction of the Minimum Independent Set - MIS (b) and Minimum Connected Dominating Set - MCDS (c) of the graph shown in (a)

the CDS. In the first phase a Maximal Independent Set (MIS) is formed. An Independent Set (IS) of a graph G is the node subset S where no two nodes in S have an edge in common. The MIS is the maximal IS, which means that it is not possible to include more nodes in S. In the second phase, the goal is to build a CDS using nodes that do not belong to the MIS. These nodes are selected in a greedy manner. At the end, the non-MIS node with the highest weight (the weight depends on the remaining energy and the degree of the node) becomes part of the CDS, as depicted in Fig. 2. Unfortunately, a CDS only preserves 1-connectivity and it is therefore very vulnerable. When fault tolerance against node failures is taken into account, the problem becomes the $kmCDS$ problem, (k-Connected m-Dominating Set). The requirement of $k - connectivity$ guarantees that between any pair of dominators there exist at least k different paths, and the $m - domination$ guarantees that each dominatee is connected with m dominators. Wu & Li (2008) propose a distributed algorithm for this problem with time complexity $O((m + \Delta) \cdot Diam)$, where Δ is the maximum node degree and $Diam$ is the diameter of the network (the length of the longest shortest path between any pair of nodes in the graph). Li (2008) assumes that the MCDS nodes are aligned according to a strip-based deployment pattern, as in Fig. 3 where the nodes are deployed in straight lines. The difference with a grid pattern is that the odd lines are horizontally shifted by a given distance in relation to the even lines. This pattern is shown to be a near-optimal solution of MCDS for an infinite network in terms of space. Because a WSN is a finite network, the spacing parameter in this pattern and consequently the number of nodes needs to be adapted. The optimization problem aims to minimize the number of nodes in the strip-based pattern such that the areas, defined by the node's transmission range, of three neighboring nodes in this pattern intersect each other (see Fig. 3). The solution of this problem gives the positions of the CDS. Implementation in a real scenario is easier. Assuming a given finite area with the sensor nodes uniformly deployed in it, for every position determined by the algorithm the closest sensor in the network will be selected for belonging to the CDS. The distributed approach requires

Fig. 3. A strip-based pattern (redrawn from (Li, 2008)).

that nodes exchange certain information, such as the distance from the ideal positions and the number of neighbors that they cover, in order to make a decision regarding membership of the CDS. Nonetheless, the problem of finding the MCDS becomes more complex for dynamic or mobile networks, and this question is still open.

Up to now we have taken "neighbors" to refer to those nodes that are reachable if a node transmits with a given power. In (Liu et al., 2010; Ma et al., 2008) the authors also take into account the existence of *lossy links*. A lossy link has an additional parameter representing the probability of a successful transmission over the link. Topology control algorithms that consider these links are known as *opportunistic* algorithms. The related problem in (Ma et al., 2008) is to minimize the number of hops between a node in the network and the sink while guaranteeing that the path utility (the utility used is a metric reflecting the expected number of packet transmission required to successfully deliver a packet) falls within in a given interval. The distributed approach requires that a node knows the utility value and the IDs of its $2 - hop$ neighbors and that it decides whether or not to act as a relay node. (Liu et al., 2010), on the other hand, demonstrated that the problem of finding a subnetwork of the original network (the subnetwork has to contain all the nodes but only a subset of link of the original network) which minimizes the overall energy consumption and guarantees that the *reachability coefficient* (RC) for every node-sink pair exceeds a particular threshold is NP-hard. RC is a coefficient that indicates the probability of a node being able to reach another node in the network while the respective threshold is imposed by application requirement. When calculating the RC for two nodes that are connected by a path, the RC will be equal to the

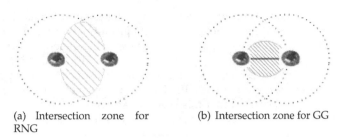

(a) Intersection zone for RNG

(b) Intersection zone for GG

Fig. 4. Intersection zones for constructing Relative Neighbor (RNG) and Gabriel (GG) graphs.

mean of the RC values of the links that constitute the path. The key idea of their solution is that link-disjoint trees can be constructed, the union of which will give the subnetwork. A node will make a decision to join in some tree construction if its RC is less than a particular threshold.

3.3 Power control

Unlike the topology control problem which seeks to minimize the size of the network backbone while assuming uniform and constant power transmission, the power control problem (also referred to as the Range Assignment (RA) problem or Strong Minimum Energy Topology (SMET)) aims to fix the node's transmission power at appropriate levels. The goal here is to reduce energy consumption while preserving connectivity in the network. Different methods proposed in the literature for solving this problem are discussed in this subsection. We present some extended versions which add new constraints to this problem with respect to i) throughput, ii) traffic and iii) reliability.

SMET has been shown by Cheng et al. (2003) to be an NP-hard optimization problem. To tackle the problem they propose two heuristics: Minimum Spanning Tree (MST), where power is assigned to nodes such that they can reach the farthest children in the MST, and Incremental Power (IP). In the IP heuristic the power of the node is allocated in a greedy manner. The heuristic begins with an empty set of nodes, to which it then adds a node chosen randomly from the network. This node adjusts its power to reach its closest neighbor. Further, each member of the set tries to increase its power to include another node, but the only member to succeed will be the one that expends the least energy in achieving this end. The algorithm stops when all the nodes are included in the set. Since transmitting with the same power can lead to energy waste, some methods based on computational geometry, such as Relative Neighbor Graph (RNG) (Wan et al., 2001), Gabriel Graph (GG) (Ke et al., 2009), Yao graph or Voronoi Diagram have been put forward to determine the "best neighborhood". In these methods two nodes can be neighbors if there are no other nodes in the zone of intersection. The main difference between them is the way that they define this intersection zone. Fig. 4 shows how the intersection zone is constructed in RNG and GG. The idea behind computational geometry implementations is that the energy cost of transmitting directly to some nodes would be less than the cost of using any other relaying scheme to reach them, and so it is worthwhile to use certain methods to discover a node's best neighbors. In many cases the node can reduce its energy so as to be able to reach only its best neighbors. Many algorithms proposed to construct these graphs are centralized, but there also exist distributed versions (Li et al., 2002). A memetic algorithm is proposed by Konstantinidis et al. (2007)

to solve the SMET problem. In reality, the difficulty of applying this kind of algorithm is modeling the problem according to the algorithm's logic, and deciding for example how to define a chromosome, how to implement crossover, how to handle population diversity, etc. The solution to the SMET problem takes the form of an array of positive integers, in which the elements of the array correspond to the power levels assigned to each node, and the respective indexes correspond to the node ID. In Fig. 5 we have 5 sensor nodes which are transmitting with a given power. From this scenario an array of 5 elements is constructed which contains the power values ordered by node ID. This array alternatively represents an

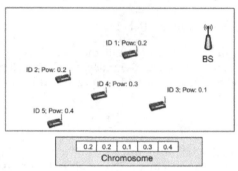

Fig. 5. Chromosome Construction

individual, a chromosome, or a solution, depending on the point of view. The objective of the SMET problem is given by the fitness function defined by the sum of the powers assigned to each node. The first phase of the algorithm proceeds by initializing a random population. It then applies a local search to check the feasibility of the solutions, modifies them in order to obtain feasible ones, and improves the solutions by reducing the assigned power if it is possible. In the second phase a genetic algorithm is applied which involves the crossover of the selected individuals and the mutation for maintaining population diversity. Finally the best individuals from each generation are generated. The procedure is repeated until the solution cannot be further improved.

Lately, this problem has been extended to take some other important parameters into account. The problem of maximizing *the throughput* using topology control is discussed in (Tao et al., 2010). Assuming that the WSN is presented through an RNG (or a GG), their algorithm adjusts the intersection zone between two neighbor nodes in the respective graph (the intersection zone between two neighbors is depicted in Fig. 4) such that the throughput is maximized. They show that if the area of the intersection zone between two neighboring nodes changes in a given interval, the network will preserve the connectivity and energy efficiency properties. Their solution is based on mathematical analysis and a complex equation is derived to find the optimal solution which guarantees the maximal throughput. The equation takes as inputs the node density and the expected throughput of the network. In Gogu et al. (2010), on the other hand, there is a discussion concerning the problem of transmission range assignment and optimal deployment to reduce the energy consumption while taking node traffic into account. The solution is based on dynamic programming methods and it gives the optimal number of sensor nodes and their transmission ranges for a linear network operating under different traffic scenarios. This work also includes an extension to the multihop network case with aggregation (Fig. 6). Hence for a given random deployment of sensors (the blue points in figure), the algorithm calculates the number of nodes that will be in charge for aggregating and relaying the data towards the base station (the red points), their location, and the respective transmission range. Valli & Dananjayan (2008) discuss the problem of topology control to

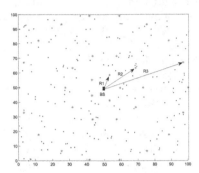

Fig. 6. Optimal position of sensors (red points) in a random deployment.

maximize *network reliability* measured by the bit error rate (BER). They model the problem as a game where a node in the network represents a player. Based on some local information a node calculates the utility function which depends on the link's BER. In every iteration each node will try to optimize this function in a non-cooperative way until the system reaches Nash equilibrium. Another approach is adopted by (Yang & Cai, 2008) to deal with *QoS requirements*. Residual energy, end-to-end delay and link loss ratio are the QoS parameters considered. The question is how to allocate the power values to the nodes such that the energy consumption is minimized, the network is connected and the above QoS requirements are met. The solution proposed is a distributed heuristic based on the minimum spanning tree (MST), where the link metric is a function of delay and packet loss ratio. When this tree is constructed, a node adjusts its power simply so as to be able to reach its parent.

3.4 Medium access strategies

In this subsection we are looking at a node's strategies for accessing the medium. These strategies govern the coordination between the nodes in the network in order for them to access the medium and perform successful transmissions. Most of the work related to medium access strategies in WSN is related to the two main approaches, which are, first, scheduled and secondly, random access/contention-based (Ye & Heidemann, 2003). TDMA (Time Division Multiple Access) is one of the common mechanisms falling under the scheduled approaches, whereas CSMA (Carrier Sense Multiple Access) and derivatives are the most commonly-used methods based on channel contention. Other solutions, more common in cellular networks, but also used by the WSN community, are FDMA (Frequency Division Multiple Access) and CDMA (Code Division Multiple Access). The TDMA, FDMA and CDMA mechanisms are employed in WSN to ensure a collision-free medium access. In this subsection we present the basic problem related to each of them. We then describe three extended versions of TDMA relating to i) connectivity, ii) traffic and iii) delay. For FDMA the extended constraint is throughput. Regarding CDMA, we discuss the problems related to joint use of CDMA with TDMA or FDMA.

Under the scheduled approach the basic problem is to obtain a slot allocation for all nodes in the network using the smallest possible number of slots such that k-hop neighbor nodes (where k is a positive integer usually equal to 2) are not allocated to the same time slot. The respective optimization problem is the chromatic graph optimization problem, which aims to minimize the number of colors used to color the nodes such that two neighbor elements do not use the same color. This problem is addressed in several works that have put forward a number of distributed algorithms (Al-Khdour & Baroudi, 2010; Gandham et al., 2005; Kawano & Miyazaki, 2009; Sridharan & Krishnamachari, 2004). In (Sridharan & Krishnamachari, 2004) slot allocation uses the logic of a breadth-first search algorithm where the first node which allocates the slot is the root of the tree (the sink). Once a node is selected it continues the operation of slot allocation based on the information from its neighbors. Gandham et al. (2005) discuss the edge-coloring problem, where two edges incident on the same node cannot be assigned to the same time slot. They propose a greedy heuristic whose first step involves coloring the edges and whose second step proposes a strategy to map the colors to the time slots. The second step uses the edge orientation to avoid the hidden and exposed node terminal problem. A simple example is shown in Fig. 7. The process begins with node 6 (the node with the largest ID), which picks a color from a set of colors and broadcasts this information to its neighbors. On reception of the information node 5 picks a different color, and so on. This process continues until all the nodes have colored their edges. Then, edge orientation is applied to the edges with the same color. So, for instance, in Fig. 7(c) let us imagine the case where node 4 transmits to 6. Because of the node 4 transmission, the level of interference may be sufficiently high to corrupt the transmission of the link $(2,3)$

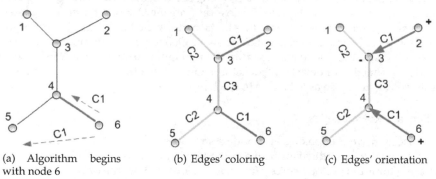

(a) Algorithm begins with node 6   (b) Edges' coloring   (c) Edges' orientation

Fig. 7. Edge coloring algorithm

The same problem is reexamined by (Al-Khdour & Baroudi, 2010), under the assumption that nodes can communicate with different frequencies. Nowadays radio chips support multichannel transceivers which can help to reduce the number of required time slots in a TDMA frame. The distributed heuristic algorithm proposed in this work is based on solving the TDMA problem in a tree structure. The base station collects the information from its children to calculate how many slots are needed (e.g. 3 slots are required in Fig. 8(a)). Next, every parent allocates a time slot to its children 8(b). Each branch of the tree will use a different channel (the frequencies can be repeated in space), whereas the nodes in one branch will transmit in different slots.

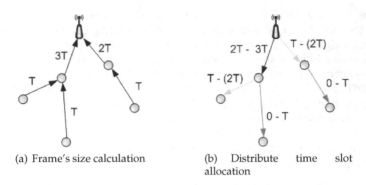

(a) Frame's size calculation

(b) Distribute time slot allocation

Fig. 8. Time slot allocation based on a tree structure

In other versions of the problem the scheduling solution must satisfy certain requirements such as connectivity, data rates and delay. Kedad et al. (2010) formulate the problem as follows: construct a frame with the minimum number of time slots such that at each time slot the activated links are not in conflict, and form a strongly-connected graph. The second constraint ensures that a node will be able to send a data packet to any other node in the network through the activated links. The links will be in conflict if they have the same transmitting or receiving node, or if the transmitting node of one link is the same as the receiving node of the other link. They show that this problem is NP-hard and propose two approximation algorithms. In (Ergen & Varaiya, 2010) the problem is to find an available slot allocation with minimum frame length, taking into account the quantity of data that a node needs to transmit. Notice that a link can be scheduled more than once in a time frame to satisfy the node data rates, which is the main difference with the basic version. Wang et al. (2007) formulate a multi-objective optimization problem. The question is to find a time slot allocation that satisfies i) the data delivery delay and ii) the node energy constraint. Here, not only are the transmitting and receiving energies taken into account, but also the energy consumed in switching between sleep and active modes. The two selected objectives contradict each other, since the energy objective seeks to maximize the number of nodes that are turned off, which in turn increases the delay. The trade-off between energy and delay is solved using the particle swarm optimization approach. This example gives a meaningful illustration of interdependence between problems coming from different layers. We have here a scheduling problem combined with a routing one in the sense that the latter one is responsible for the delay.

While TDMA-based approaches schedule transmissions sequentially over time, FDMA-based approaches permit multiple concurrent transmissions between neighboring nodes by allocating different channels/frequencies to them. Sensors in the network can thus tune their operating frequency over different channels to avoid interference and packet collisions in the network. One of the advantages of FDMA is the improvement of network throughput and packet transfer delay. In FDMA the problem can also be modeled as a graph-coloring problem, given that no two adjacent nodes are allowed to use the same channel.

Yu et al. (2010) show that the problem of assigning the channels such that *the interference* is minimized is NP-hard. They model the problem as a game where every node is a player and the interference is the objective to be minimized. Their algorithm assumes that routing is based on the tree structure. In each iteration an intermediate node selects its own channel so as to cause the least possible interference for its neighboring nodes. The interference is calculated using local data that include the number of interfering parents in different branches existing in their neighborhood, their respective numbers of children, and whether or not these children are leaves within the tree structure. Notice that neighbors of a given node can belong to different branches and have different roles, parents or leaves. Based on an empirical study, Wu et al. (2008) find it more appropriate for a WSN to communicate using a single channel, but they suggest harnessing channel diversity by spreading the frequencies in space. They therefore propose a node-disjoint tree structure where every branch (subtree) communicates via a given channel. The objective here is to divide the network into multiple disjoint subtrees such that the interference between them is minimized. They show that the problem is NP-hard and propose a greedy heuristic.

CDMA spreads the baseband signal using different *Pseudo Noise (PN) codes* to enable multiple concurrent transmissions. In WSN, a PN code may be implemented as an attribute in the packet header (nodes simply need to check whether the code in an incoming packet matches their own set of codes) in order to reduce the complexity of modulation and decoding techniques in comparison to CDMA implementations using other technologies. Optimization problems relating to code allocation in CDMA are slightly different from those relating to time or channel allocation. For instance, in CDMA it is possible that two neighboring nodes share the same code but only one can use the code for transmitting while the other node can use it for receiving. The optimization problem may require that no two adjacent directed links have the same code. The difference between WSN and other wireless CDMA networks is not really to be found in the problem of code allocation, but in the CDMA concept itself. CDMA codes are not completely orthogonal. The high density of sensors in the network makes the problem of interference in concurrent transmission a very serious one. High interference causes problem for receiver nodes because they cannot 'understand' the signal addressed to them. In the literature the pure CDMA problem is addressed simultaneously with the channel and slot allocation problems. The problem of *channel and code allocation* to reduce interference is discussed in (Liu et al., 2006). Their distributed solution is a heuristic which tries to solve first the problem of channel allocation and subsequently the code allocation one. When CDMA is combined with *scheduling*, Chen et al. (2006) looks for a feasible schedule for all the nodes in the network, together with their respective PN codes such that there is no interference (or the interference falls below a given threshold) in any time slot and the total energy consumption is minimized.

3.5 Duty cycle

The node duty cycle is determined by its activity and sleep periods. During the sleep periods the sensor nodes do not consume energy, and so short activity periods mean energy savings. However, this has to be scheduled, because nodes can communicate with each other only during the activity periods. The set of active nodes in the network at a given moment must satisfy certain requirements, the most important being connectivity and coverage. In the first paragraph below we discuss the problem of node scheduling with a connectivity constraint. In the second paragraph coverage is taken into account and two additional constraints are introduced, namely i) life dependency between sensors and ii) connectivity.

(Nieberg, 2006) models the node duty cycle with a connectivity constraint as the MCDS problem. He also proposes a distributed algorithm for finding this set of nodes. Here it is assumed that the network is very dense and nodes are close to each other such that a large number of nodes can become passive while the remaining nodes continue to ensure a connected structure. The active nodes correspond precisely to the CDS. According to the algorithm some nodes will have a special role: those nodes that form a Maximal Independent Set perform the role of *anchors*, and nodes used to connect anchor nodes perform the role of *bridges*. Nieberg (2006) shows that the set of anchor and bridge nodes forms the CDS. The initialization phase has self-organizing properties. Each node will try to get an active time slot according to the TDMA scheme. Then, any other node that enters into the network needs to decide locally whether or not it will be active (either as a bridge or as an anchor). The decision is based on the information provided by the neighbor nodes. This information includes the neighbor node ID, a list of all time slots showing the slots occupied by them and their respective neighbors, their role as an active node, and some synchronization information. When a node observes that there are less than two anchor nodes in its neighborhood for a given time slot, it becomes an anchor otherwise it seeks for the existence of bridge nodes. If it finds that any pair of anchor nodes are connected with bridges, it becomes passive.

The node duty cycle is also related to coverage requirements. Because monitoring is one of the main objectives of a WSN, the active nodes have to guarantee that a set of given targets will be monitored throughout the lifetime of the WSN. The problem is to group the nodes such that i) each group (known as a cover) is able to cover the targets and ii) the groups form disjoint sets of nodes in order to maximize the WSN lifetime. Usually a redundant sensor network is considered in this case. This question has elements of both a coverage problem (targets which need to be covered) and a scheduling problem. Only the nodes belonging to a cover are to be activated, while the others are to be put to sleep, and the covers are to be activated in a sequential manner. Cardei & Du (2005) have shown that this problem is NP-hard. In (Rossi et al., 2010), the problem is modeled as a linear program whose aim is to maximize the sum of the different covers' lifetimes, the constraint being that the total duration of a nodes' activity periods does not exceed the lifetime of its battery. The problem is solved using the column generation method. Aioffi et al. (2007) model the problem as the weighted set cover problem (WSCP). Given n sets $(S_1, S_2, ...S_n)$ formed from elements of a universal set denoted the US, together with their associated activation costs, WSCP seeks to find a subset of these sets such that the sum of the activation costs is minimized, and whose union corresponds to the US. The set of the targets in the network problem is modeled as the US, the sets S_i represent the set of the target covered by sensor i, and the cost of S_i is the inverse of energy for the sensor i. The problem is solved off-line and the results are fed into the sink. When the mobile sink gathers data from the nodes, it also indicates to them whether they will be active in the following period. This method is used particularly for this case because the number of possible combinations is exponential (the number of constraints is very small) and it can achieve faster convergence.

(Dhawan & Prasad, 2009) remove the constraint of disjoint covers. If a node is included in two or more cover sets, then its energy capacity will influence the life of these sets. They therefore propose a solution based on the construction of a life dependency (LD) graph. In this graph covers are represented by vertices, linked by an edge if they share the same sensors. The LD graph is introduced into the problem in order to identify the covers having the least impact on the other covers. Their distributed approach adds a communication cost between neighboring nodes which need to exchange information such as the remaining energy and the region (area or targets) they can cover. A further cost is added, corresponding to the processing of this information and to making a decision. Every sensor thus needs to construct an LD graph based on its local information and to identify the cover with the smallest impact in order to be part of it. Finally, there is also a communication cost corresponding to the negotiation phase where nodes attempt to obtain a stable solution. In (Cardei & Cardei, 2008; Zou & Chakrabarty, 2005) the same problem is discussed and an additional constraint imposed: each set is required to be connected with the base station. In (Cardei & Cardei, 2008) the problem is formulated as Integer Linear Programming. It is first centrally solved using ILOG CPLEX, and then via a distributed approach. In the distributed case each node needs to know not only its own coordinates but also those of the given targets and base station. The initialization phase has a considerable communication cost resulting from exchanging the list of targets that the two-hop neighbors cover, the status of every node, and the synchronization message. This initialization phase includes the creation of the cover sets, while the subsequent phase finds the relaying nodes for connecting the cover with the base station (one node in the cover constructs a spanning tree that includes the target set and the BS).

4. Routing

Data transmission in WSNs, also referred to as the routing problem, is one of the most widely studied problems in WSN. Different to the previous section, we focus here on the main proposed models and give some analysis on their use. The models and methods used for solving routing problems in WSN can be roughly divided in two main groups. The first group includes related shortest and spanning tree models, while the second group is centered around flow models and comprises a range of different minimum cost/maximum multicommodity flow models. While abundant work relating to such problems exists for wired networks, some new challenges have appeared for wireless networks, and especially for WSNs. The nature of some of these problems can change quite radically when they are placed in a WSN context and new requirements are introduced. These requirements include sensors' energy constraints, the interference caused by the broadcast nature of transmissions over wireless links, as well as data compression, aggregation and processing constraints. For instance, in traditional formulations of the network flow problem, link capacity is a strong constraint, while in WSN this constraint is frequently supplanted by the node energy constraint. Another important difference between these two paradigms is the inclusion of the dynamic topology models and the need for distributed solutions for wireless sensor networks.

4.1 Shortest Path and Spanning Tree based models

Shortest Path Tree (SPT) and Minimum Spanning Tree (MST) remain widely used models for routing design, even in WSNs. The goal of a SPT is to find a path of minimum cost from a specified *source node* to another specified *sink node*, assuming that each edge has an associated cost. In the WSN context the edge cost usually represents the power that would be consumed by the transmitting node when sending a packet to the node at the opposite end of the edge. Distributed routing algorithms based on Dijkstra, Bellman-Ford or Chandy-Misra's distributed algorithms can thus be employed (Rodoplu & H., 1999; Yilmaz & Erciyes, 2010). One of the disadvantages of SPT is the unbalanced load between the sensors and the disparity in the energy used by them that such methods can lead to. To overcome this problem, different strategies are proposed. In (Yilmaz & Erciyes, 2010) every node can regenerate a path when a fault occurs or available energy is depleted. Other works consider edge cost to be a combination of several metrics such as residual energy, buffer size, or the number of neighboring nodes.

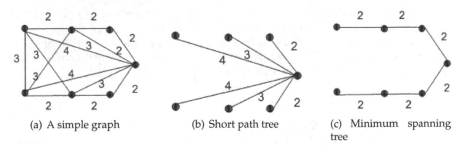

(a) A simple graph   (b) Short path tree   (c) Minimum spanning tree

Fig. 9. Shortest path (b) and minimum spanning tree (c) for the graph shown in (a)

Going further, WSN brings new constraints which may modify the nature of the problem. For instance, many applications of WSN require that the intermediate or relay nodes aggregate the data, while the criterion used is minimizing energy consumption. For (Cristescu et al., 2006) the joint problem of data aggregation and routing is NP-hard, and their heuristic combines an MST with an SPT. Normally, in cases where there is a high aggregation coefficient, the amount of traffic increases slightly from the source to the sinks, and an MST is a good compromise. However, where the aggregation coefficient is low, routes need to be found that minimize the number of hops, and therefore an SPT should be constructed. MST is the tree structure which minimizes the sum of edge costs, and the problem is polynomial. The difference between a shortest path tree and a minimum spanning tree is shown in Fig. 9.

Minimizing the total energy consumption is, however, not enough, since some nodes deplete their energy faster than others and may cause network partition. To balance the energy consumption, one strategy is to minimize the maximum energy consumption of the nodes. This problem has been modeled by (Gagarin et al., 2009) as the minimum degree spanning tree (MDST), which is an NP-hard optimization problem. Variations of this problem are encountered in the literature, in (Erciyes et al., 2008; Huang et al., 2006). A joint routing and data aggregation problem is also discussed in (Karaki et al., 2009) for a two-tier network, and some heuristic algorithms such as GA and greedy are proposed. From a distributed perspective, adapted versions of Prim's and Kruskal's algorithms have been proposed in (Attarde et al., 2010). In the distributed versions of SPT a node need only communicate to its neighbors information concerning the cost of links. Each node decides to communicate with the node that provides the minimal cost to the base station. An ACK mechanism is needed to dictate the end of the process. It may be remarked here that almost all the above cited models lead to single path routing schemes. They have the great advantage of being simple from an implementation point of view, while their main drawback is their difficulty in embracing additional requirements, energy consumption in particular. We now present some flow-based models that can model such requirements in a suitable way.

4.2 Flow-based models

The need to include energy/capacity constraints leads naturally to the use of flow models. Particularly for the WSN, routing problems are formulated as MultiCommodity Flow Problems (MCFPs). The commodity is a source-destination pair, and we are faced with an MFCP whenever several commodities share the network resources. In an MCFP the commodities will have different sources and/or destinations, but they are bound together insofar as they share the same link capacities. Regarding commodities, a WSN gives rise to either single-sink or multi-sink models, and in the case of single-sink models all commodities will have the same extremity, namely the base station. In the following subsection 4.2.1 we discuss some basic versions of flow models used for routing path calculation in WSN. Then, in subsection 4.2.2 some further extended routing problems are presented.

4.2.1 Conventional flow models in WSN

A standard flow problem in WSN (regardless of whether it is a multicommodity flow problem) includes two type of constraints, namely the flow conservation constraint and the energy constraint.

$$\sum_{j \in N_i} x_{ij}(t) = \sum_{j \in N_i} x_{ji}(t) + y_i(t) \qquad \forall i \in N, \forall j \in N_i, \forall t \in T, \tag{5}$$

$$\sum_{t \in T} \sum_{j \in N_i} x_{ij}(T) * e_{ij} \leq E_i \qquad \forall i \in N, \forall j \in N_i, \tag{6}$$

where t (respectively T) is a time instance (respectively the network lifetime), N the set of sensors, N_i the set of neighboring nodes of i, x_{ij} the flow over the edge ij (that is to say the data transmitted over this link), y_i the data generated by node i, e_{ij} the energy consumed in transmitting a unit flow and E_i the initial energy of the sensor. The flow conservation constraint, Equation (5), shows that the total amount of flow that a sensor receives plus the amount of data that it generates is equal to the amount of information that it transmits. The second constraint given in Equation (6) is the capacity constraint, which is related to energy. This constraint implies that the energy consumed by a sensor for transmitting the flow throughout the lifetime of the network must be less than its initial energy. In standard network flow problems this constraint is usually related to link capacity.

One of the first works to formulate this problem in terms of Integer Linear Programming is to be found in (Chang & Tassiulas, 2004). The flow is represented here by the number of packets and the transmission energy is calculated based on the distance between the nodes (and hence assuming a power control mechanism). The optimal solution of this problem gives an upper bound for network lifetime. While the problem of lifetime or flow maximization under these constraints can be solved in polynomial time for continuous values of flow x, the integer version is shown to be strongly NP-hard in Bodlaender et al. (2010). The distributed version of this problem is discussed in (Madan & Lall, 2006), where the subgradient algorithm is used to solve the problem. At each iteration the algorithm estimates the gradient value at a given point of the objective function and determines the next point to be considered, until the optimum is reached. The distributed implementation of this algorithm requires that every node keeps track of two variables, namely the *flow rate* of every outgoing link and the *network lifetime*. These variables are updated during each iteration of the algorithm based on their previous values and the subgradient function values (also a function of flow rates and network lifetime) are calculated according the information received from neighbor nodes. Subgradient methods are also used by Rabbat & Nowak (2004) as convenient tools for designing a distributed approach in sensor networks. Another characteristic of WSNs is the data aggregation applied by nodes. This phenomenon can easily be taken into account by slightly modifying the conservation flow constraint. For instance, in Cheng et al. (2009) each node sends the maximum amount of information between the received and the generated data set as in Equation (7).

$$\sum_{j \in N_i} x_{ij} = max_{j \in N_i} \left\{ x_{ji}, y_i \right\} \qquad \forall i \in N \qquad (7)$$

The routing problem with data aggregation for lifetime maximization in a network has been formulated by Xue et al. (2005) as a concurrent multicommodity flow problem. Here the flow constraint implies that the amount of the flow commodities transmitted from a sensor node cannot be less than the sensor's data. They propose a polynomial time approximation scheme, strongly inspired by the Garg-Konemann algorithm. In outline, their algorithm is as follows: construct the shortest path between every source and the sink, initialize a cost unit flow for every node, push the maximum possible flow along the path for every commodity, update the cost of energy for every node and repeat the process.

As regards routing paths, the routing schemes can use several paths (in other words perform multipath routing), or a single path (single-path routing.) Although requiring routing via a single path would appear preferable for WSN, adding such a constraint to the mathematical model gives rise to NP-hard problems. Worth citing here are two approaches proposed for WSN that attempt to circumvent the computational burden of such models while providing simplicity in implementation. The first approach computes a solution involving multiple paths, but uses only one single path at a time. Hou et al. (2004) propose an algorithm to solve the problem in two phases. In the first phase a solution is found for the multipath routing problem. Consequently every node knows the set of the relaying nodes and the respective amount of information to send to them. In the second phase one node, according to some local rule, will select one of its relaying nodes and will transmit to it the whole amount of information to be sent in this round. The second approach, in stark contrast to the first approach just described where routing takes place from the sensors to the BS (i.e. flat routing), may be seen as hierarchical routing, in that it decomposes the data transmission into two levels and thus converges to a cluster-based scheme. Each cluster head (CH) receives the data from the nodes of its cluster and from the other CHs, and transmits this data to another CH in the direction of the BS. Bari et al. (2008) consider a two-tier heterogeneous network containing powerful relay nodes which form a connected network that can relay data to the BS. They formulate the optimization problem as follows: knowing the positions of sensors and relay (CH) nodes, how should the network be clustered in order to maximize its lifetime? A sensor is not obliged to transmit directly to the CH, and sensors may have different amounts of flow to transmit. The problem is formulated as a max-min LP. Because the decision variables can take only binary values (1 if the sensor belongs to a given cluster and 0 otherwise) and the flow rate variable corresponds to a number of bits, we are dealing with an ILP problem. The heuristics presented for this problem are centralized. Other centralized techniques for solving the clustering problem in WSN are based on Fuzzy Logic (FL) (Anno et al., 2007; Ran et al., 2010), while Mehrjoo et al. (2011) proposes genetic algorithms.

4.2.2 Enhanced flow based models

Advances in technology and the broad range of applications for WSN have given rise to new QoS requirements and made routing a more complex matter. Interference, delay and questions of reliability may all place additional constraints and lead to more elaborate and challenging versions of routing problems. All this will be in the focus of this paragraph.

Radio interference has a significant impact on the performance of WSN as it affects the functioning of both MAC and routing protocols, and directly affects the transmission capacity of links. In contrast to traditional networks where the capacity of links is determined by physical parameters only, in wireless communications radio interference strongly affects the transmission capacity of links that are located close to one another. The models we have cited above assume that the quantity of information generated is sufficiently low, or the channel capacity sufficiently high, for transmission capacity not to be an issue. But this assumption clearly does not always hold, and capacity constraints over links are sometimes unavoidable. It should be noted that IEEE 802.15.4 defines data rates of 20, 40, or 250 Kb/s for the physical layers. Channel capacity may therefore represent a strong constraint where huge amounts of data need to be transmitted, or when many sources have to transmit simultaneously. Interference needs to be taken into account because of the high bit error rates that it may cause. The capacity of wireless channels is calculated from the Shannon-Hartley formula given in Equation (8).

$$C = B \cdot \log_2 \left(1 + \frac{S}{N} \right) \tag{8}$$

where C is the channel capacity (in bits per second), B the channel bandwidth (Hz) and S/N the signal-to-noise ratio.

From the point of view of computational complexity, including this constraint in the model makes the problem NP-hard, as shown in (Jain et al., 2003). More precisely, they show that the problem of finding a maximal flow for a source-destination pair under the interference constraint is equivalent to the Minimum Independent Set problem in a graph, and therefore NP-hard.

Krishnamachari & Ordonez (2003) add the link capacity constraint to the basic version of the flow problem with the goal of maximizing the throughput or minimizing the overall energy consumption. To ensure that the solution will not generate scenarios in which the traffic load is unfair for the nodes in the network, the flow transmitted by a node has to be less than a given fraction of the total flow generated by the network. Patel et al. (2006) add the following two constraints to the basic version of the routing problem: (i) the link capacity constraint where the rate (the number of packets per unit time) at each link has to be smaller than its capacity, and (ii) the node capacity constraint where the number of packets that a node can process in a unit time has to be smaller than its given capacity. The proposed algorithm is centralized and aims to find a maximum flow with the smallest possible energy cost. It is a kind of combination of maximum flow (getting as much flow as possible from the source to the sink) and shortest path (traveling from the source to the sink with minimum cost). The problem addressed in (Xu et al., 2008) has the same structure as that found in Patel et al. (2006), but the objective is utility maximization, which is a nonlinear convex function of the transmission rate. The problem is solved using the Lagrangian method. This method attempts to decompose the problem into a number of sub-problems via a Lagrange multiplier and to solve each of them separately. In these problems it is assumed that the bandwidth B is shared between different node channels, or that the nodes use the whole bandwidth but are already scheduled in order to avoid interference.

There are two possible ways of modeling a successful transmission in the presence of interference: i) the physical context, which requires that the Signal-to-Interference and Noise Ratio (SINR) given in Equation (10) exceeds a certain threshold; ii) the protocol context, where no two neighboring nodes may transmit at the same time.

Routing under the physical interference model is more complex. Wang et al. (2011) discuss a link scheduling problem where flow capacities are satisfied and the time taken for scheduling is minimized. In this case the channel capacity is variable over time due to SINR, and its integral gives the service provided by the channel as expressed in Equation (9).

$$C_{ij}(t) = \int_0^t B \cdot log(1 + SINR_{ij}(\tau))d\tau \tag{9}$$

where $C_{ij}(t)$ is the channel service of link (i, j) during time t, and B is the channel bandwidth.

$$SINR_{ij} = \frac{\omega_{ij}(t)P_i}{\sum_{k \in V^+ / \{i\}} (\omega_{kj}(t)P_k) + N_a} \tag{10}$$

where $SINR_{ij}$ is the $SINR$ parameter for the link (i, j), ω_{ij} the gain of the fading channel for the link ij, P_i the power transmission of node i, $\omega_{kj}P_k$ measures the interference of the other links over the link (ij) and N_a is the floor noise which is a constant. The channel service calculated in each time slot is used as parameter to bound the link data rate. The problem is solved off-line using the column generation method.

Interference can be more easily modeled in a protocol context. Wang et al. (2008) study the routing problem in the presence of interference by scheduling the nodes in accordance with the TDMA approach. The constraint added for the interference implies that the sum of the number of times a link is scheduled plus the sum of the number of times that all the links in its interference zone are scheduled in the time frame has to be smaller than the frame size, as in Equation (11).

$$N(e) + \sum_{e' \in I(e)} N(e') \leq S \tag{11}$$

where $N(e)$ is the number of times that the edge e is scheduled in the time frame, $I(e)$ is the subset of links of the original graph that can be influenced from e transmissions and S is the number of time slots in the frame.

We shall now focus on how WSN takes some QoS requirements and their associated metrics into consideration. We begin with a discussion of QoS metrics and the computational complexity that they introduce. Different metrics have different composition rules. Metrics such as delay, delay jitter and cost are additive (an additive metric is a metric which obeys the additive rule, meaning that the path metric is equal to the sum of the metric links that compose the relevant path). A multiplicative metric is a metric which obeys the multiplicative rule, meaning that the path metric is equal to the product of the link metric for all the links that compose the relevant path. Metrics like reliability (the probability that the transmission was successful) can thus be seen to be multiplicative. Finally, concave metrics obey the concave rule, meaning that the path metric is equal to the minimum (or maximum) link metric for all the links that compose the relevant path. Bandwidth is an example of a concave metric. Fig. 10 illustrates the concept of multicommodity flows in a graph and QoS multipath routing with two metrics.

In (Wang & Crowcroft, 1996) it is shown that the problem of finding a path which satisfies N additive metrics, and/or K multiplicative metrics (where N and K are positive integers) is NP-hard, while it becomes polynomial when one is concave and the other additive or multiplicative.

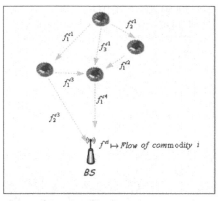

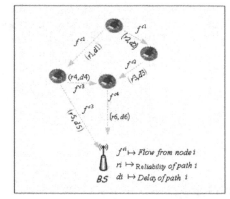

(a) multicommodity flows in a network   (b) QoS multipath routing with two metrics

Fig. 10. Multicommodity and multipath Routing

Most works dealing with QoS routing in WSN are concerned either with finding (disjoint) paths for guaranteeing network resilience (fault-tolerant network), or with finding a minimal number of paths such that QoS requirements are met. We recall that the problem of finding k disjoint paths (edge or vertex disjoint) such that the total cost of the paths is minimized has been shown in (Li et al., 1992) to be NP-hard, even for $k = 2$ in directed graphs. Heuristics therefore provide practical approaches for solving these kinds of problems. (Okdem & Karaboga, 2009) report an approach combining ACO with a tabu search. Each source node wishing to transmit data toward the BS has to launch n ants (n corresponds to the number of data packages that the source transmits). The ant's movement is based on the probabilistic decision where the heuristic value represents the estimation of the residual energy. After all the ants have completed their journey (from source to destination), each ant k deposits a quantity of pheromone equal to the inverse of the total number of nodes included in the path. This task is performed by sending ant k back to its source node following the arrival path. In this type of ACO each receiver node has to maintain a tabu list with the identities of the ants that it has encountered, enabling it to decide whether to accept the upcoming packet of ant k. Routing the information efficiently to guarantee the delay and reliability constraint is discussed in Saleem et al. (2010), who proposes a multi-agent approach for ant colony optimization (ACO). The movement of the ant is guided by the probabilistic decision rule, equation (4). The pheromone value corresponds to the end-to-end delay. The two heuristic evaluation parameters of every edge are determined by the residual energy at the extremity of the edge and its packet receive rate (PRR).

In (Bagula & Mazandu, 2008) the QoS routing problem is concerned with delay and reliability criteria. The goal is to find the smallest set of disjoint paths between a source and a destination such that both criteria are satisfied and energy consumption is minimized. Delay is a stringent metric, meaning that if the delay is not respected in any of the set paths, the packet is dropped. In contrast, the reliability of every source-destination connection obeys the multiplicative composition rule. Hence the more paths in the set, the more reliable the set will be. The problems of finding the path which minimizes the energy or the delay, or maximizes the reliability, taken separately, are solvable in polynomial time, but the problem considered in its entirety is not.

5. Open issues and concluding remarks

There are several issues in WSN which are still open or which have not been sufficiently addressed.

- Dynamicity is one of the most noticeable characteristics of WSN and also one of the biggest challenges. The term covers such phenomena as node failure, link fluctuations, node attacks and mobile nodes. Many studies in routing, coverage, scheduling or topology control have attempted to find solutions where these events occur, but including them in optimization problem models remains a challenge.

- We consider that scalability is an important issue which is frequently neglected when solution methods are proposed. The eventually changes in network dimensioning may sometimes require to resolve the problem or to sufficiently increase the computation time. We observe this particularly in relation to issues related to multi-sink/multicommodity design and network cross layer design.

- With respect to coverage problems, there are several potential directions that have not been fully explored. These include solving the deployment problem in the presence of obstacles, taking into account the restrictions for node placement and $3D$ deployments. In routing and topology control, cooperative decision-making strategies and opportunistic approaches also need to be modeled and examined in optimization problems, since in both areas some of the problems discussed here have been successfully addressed through opportunistic approach. But not many theoretical works have been undertaken in relation to this paradigm. Many questions remain open. For instance, in what scenarios should an opportunistic approach be favored over other approaches? How close is an opportunistic approach solution likely to be to the optimal solution? Routing in opportunistic networks [1] adopts a people-centric approach to model the network semantics (Verdone & Fabri, 2010). This routing group is classified as sociability-based routing and has been modeled in (Yoneki et al., 2007) based on human behavior characteristics. They propose a Socio-Aware Overlay (multi-point event dissemination using an overlay constructed by closeness centrality nodes in communities) for publish/subscribe communication. It is not clear whether these strategies might be appropriate for WSN.

- Another crucial issue is the difference that still exists between theoretical studies and practical implementations in WSN. Some theoretical studies have already presented models for cross-layer design, together with corresponding solutions. But many of them remain centralized and require off-line computation. We remark that in some mathematical formulations the variables are considered continuous, despite the discontinuous nature of the corresponding events such as power transmission and flow. On the other hand, algorithms or protocols implemented in real hardware or tested in simulations do not address cross-layer design. They aim at distributed and on-line computations and handle mostly simplified problems. Moreover, in these works the analyses that might yield an optimal solution are neglected, and it is difficult to grasp the problem complexity and to know whether there is room for further improvement. Combining these two approaches is far from straightforward and calls for substantial work. We see as a primary concern in

---

[1] Examples of opportunistic networks are Delay Tolerant Networks (DTN) (Pelusi et al., 2006.) or Pocket Switched Networks, VANETs, networks composed of devices such as MP3 players, mobile telephones and PDAs which can communicate with each other by Bluetooth or Wi-Fi to share data, or even wireless sensor networks which can send data using technologies such as GSM/UMTS, WiFi, etc.

this context the development of optimization tools and dedicated software to bridge the gap between optimization methods and their practical implementation in WSN.

- Finally, we consider that uncertainty has received very little attention until now. Nonetheless, uncertainty is an important characteristic inherent in the nature of WSNs, and is related to different aspects such as event detection, sensor location and data delivery. Some attempts to model these situations use probabilities associated with these different kinds of events. The main difficulties in taking the uncertainty of WSNs into account are twofold. First, measuring the distribution of events is not an easy task and is both environment- and application-dependent. Secondly, despite recent advances in robust optimization[2] tackling probabilistic optimization problems is not for the faint-hearted.

To conclude, wireless sensor networks represent an attractive research area due to several factors as the resource-constrained nature of sensor nodes, interference, data aggregation, power consumption model and the wide range of both commercial and military applications that this technology offers. Successful network design and deployment include understanding and modeling several problems related to these factors, which ultimately determine the available range and data rate of a WSN, as well as cost and battery lifetime. Therefore this study, intended to researchers and graduate students in computer science and fields related to operations research, information technology and applied mathematics, gives some highlights on a number of representative network problems in WSN and focuses on their respective optimization problems.

6. References

Aioffi, W., Mateus, G. & Quintao, F. (2007). Optimization issues and algorithms for WSNs with mobile sink, *Intern. Netw. Opt. Conf.* pp. 1–6.

Al-Khdour, T. & Baroudi, U. (2010). An energy-efficient distributed schedule-based communication protocol for periodic wireless sensor networks, *Arab. jour. for sci. and eng.* 35: 155–168.

Anno, J., Barolli, L., Xhafa, F. & Durresi, A. (2007). A cluster head selection method for wireless sensor networks based on fuzzy logic, *IEEE TENCON* pp. 1–4.

Attarde, S. A., Ragha, L. L. & Dhamal, S. K. (2010). An enhanced spanning tree topology for wireless sensor networks, *Int. Journal of Comp App.* 1: 46–51.

Averbakh, I. & Berman, O. (1997). Minimax regret p-center location on a network with demand uncertainty, *Elsevier Science* 5: 247–254.

Bagula, A. B. & Mazandu, K. G. (2008). Energy constrained multipath routing in wireless sensor networks, *Proc. of Ubiquitous Intell. and Comp.* pp. 453–467.

Bari, A., Jaekel, A. & Bandyopadhyay, S. (2008). Clustering strategies for improving the lifetime of two-tiered sensor networks, *Comp. Comm.* 31: 3451–3459.

Bellman, R. (1957). *Dynamic Programming*, Princeton University Press.

Bertsimas, D. & Sim, M. (2004). The price of robustness, *Operations Research*, 1: 35–53.

Bodlaender, H., Tan, R. B., van Dijk, T. & van Leeuwen, J. (2010). Integer maximum flow in wireless sensor networks with energy constraint, *Technical report*, Utrecht University.

Cardei, I. & Cardei, M. (2008). Energy-efficient connected-coverage in wireless sensor networks, *International Journal of Sensor Networks* 3: 201–210.

---

[2] Following the works of Bertsimas & Sim (2004), who showed how to model a stochastic optimization problem as a Linear Program under weak conditions, robust optimization has been intensively investigated.

Cardei, M. & Du, D. (2005). Improving wireless sensor network lifetime through power aware organization, *Wireless Networks* 11: 333–340.

Cavalier, T. M., Conner, W., Castillo, E. & Brown, S. (2007). A heuristic algorithm for minimax sensor location in the plane, *Europian Journal of Operational Research* pp. 42–55.

Chang, H. & Tassiulas, L. (2004). Maximum lifetime routing in wireless sensor networks, *IEEE trans. on Netw.* 12: 609–619.

Chen, M., Oh, C. & Yener, A. (2006). Efficient scheduling for delay constrained multi-rate CDMA systems, *Spread Spectrum Techniques and Applications* pp. 371–375.

Cheng, M., Gong, X. & Cai, L. (2009). Joint routing and link rate allocation under bandwidth and energy constraints in sensor networks, *IEEE Trans. on Wir. Comm.* 8: 3770 – 3779.

Cheng, X., Narahari, B., Simha, R., Cheng, M. X. & Liu, D. (2003). Strong minimum energy topology in WSNs: NP-Completeness and heuristics, *IEEE Trans. on Mob. Comp.* 2: 248 – 256.

Cristescu, R., Lozano, B., Vetterli, M. & Wattenhofer, R. (2006). Network correlated data gathering with explicit communication: NP-Completeness and algorithms, *Networking, IEEE/ACM* 14: 41–54.

Dantzig, G. B. (1963). *Linear Programming and Extensions*, Princeton.

Dhawan, A. & Prasad, S. K. (2009). A distributed algorithmic framework for coverage problems in wireless sensor networks, *Proc. of Parallel and Distrib. Proc.* pp. 18–25.

Dorigo, M., Maniezzo, V. & Colorni, A. (1996). The ant system: Optimization by a colony of cooperating agents, *IEEE Trans. on System Man, and Cybernetics-Part B* 26: 29–41.

Duttagupta, A., Bishnu, A. & Sengupta, I. (2008). Maximal breach in WSNs: Geometric Characterization and Algorithms, *Algosensors* pp. 126 –137.

Efrat, A., Peled, S. & Mitchel, J. (2005). Approximation algorithms for two optimal location problems in sensor networks, *Broadband Networks* 1: 714–723.

Erciyes, K., Ozsoyeller, D. & Dagdeviren, O. (2008). Distributed algorithms to form cluster based spanning trees in wireless sensor networks, *Proc. of Computer Science* pp. 519–528.

Ergen, S. & Varaiya, P. (2010). TDMA scheduling algorithms for WSN, *Wireless Networks* 16: 985 – 997.

Fidanova, S., Marinov, P. & Alba, E. (2010). ACO for optimal sensor layout, *Proceeding of International Conference on Evaluationary Computation* pp. 5–9.

Gagarin, A., Hussain, S. & T., Y. L. (2009). Distributed search for balanced energy consumption spanning trees in wireless sensor networks, *Adv. Inf. Net. and App. Work.* pp. 975–982.

Gandham, S., Dawande, M. & Prakash, R. (2005). Link scheduling in sensor networks: distributed edge coloring revisited, *Infocom* 4: 2492 – 2501.

Glover, F. (1989). Tabu search - part i, *ORSA Journal on Computing* 1: 190–206.

Gogu, A., Nace, D. & Challal, Y. (2010). A note on joint optimal transmission range assignment and deployment for wireless sensor networks, *IEEE Networks* pp. 1–6.

Holland, J. (1975). *Adaptation in natural and artificial systems*, University of Michigan Press.

Hou, Y., Shi, Y., Pan, J. & Midkiff, S. (2004). Lifetime-optimal data routing inwireless sensor networks without flow splitting, *Workshop on Broadband Advanced Sensor Networks* .

Huang, G., Li, X. & He, J. (2006). Dynamic minimal spanning tree routing protocol for large wireless sensor networks, *IProc. of Indust. Electr. and Applic.* pp. 1–5.

Jain, K., Padhye, J., Padmanabhan, V. & Qiu, L. (2003). Impact of interference on multi-hop wireless network performance, *MobiCom, ACM* pp. 66 – 80.

Karaki, J., Ul-Mustafa, R. & Kamal, A. (2009). Data aggregation and routing in WSN : Optimal and heuristic algorithms, *Computer networks* pp. 945–960.

Karmarkar, N. (1984). A new polynomial-time algorithm for linear programming, *Combinatorica* 4: 373–395.

Kawano, R. & Miyazaki, T. (2009). Distributed data aggregation in multi-sink sensor networks using a graph coloring algorithm, *Proc. of Adv. Inf. Netw. and Applic.* pp. 906–912.

Ke, W., Liqiang, W., Shiyu, C. & Song, Q. (2009). An energy-saving algorithm of WSN based on Gabriel graph, *Wir. Comm., Netw. and Mob. Comp.* pp. 1–4.

Kedad, S., Pasqual, F. & Fouilhoux, P. (2010). Ordonnancement de paquets dans les réseaux sans fil, *In Proc. of ROADEF* .

Kennedy, J. & Eberhart, R. (1995). Particle swarm optimization, *Proc. of IEEE Int. Conf. on Neural Netw.* 4: 1942 – 1948.

Konstantinidis, A., Yang, K., Chen, H. & Zhang, Q. (2007). Energy-aware topology control for WSN using memetic algorithms, *Computer Communications* pp. 2573–2764.

Krishnamachari, B. & Ordonez, F. (2003). Analysis of energy-efficient, fair routing in wireless sensor networks through non-linear optimization, *IEEE Vehicular Technology Conference* 5.

Kuhn, H. W.; Tucker, A. W. (1951). Nonlinear programming, *Proc. of 2nd Berkeley Symposium.* pp. 481–492.

Li, C., McCormick, S. & Simchi-Levi, D. (1992). Finding disjoint paths with different path-costs: Complexity and algorithms, *Networks* 22: 653–667.

Li, J. (2008). *Optimization Problems in Wireless Sensor and Passive Optical Networks.*, PhD thesis, The University of Melbourne, Australia.

Li, X., Calinescu, Y. & Wan, G. (2002). Distributed construction of a planar spanner and routing for ad hoc wireless networks., *In: Proc. of IEEE Infocom* pp. 1268–1277.

Liu, B., Otis, B., Chou, C. & Jha, S. (2006). A novel multi-channel CDMA system for wireless sensor networks, *Sensor Networks, ACM* 5: 1–30.

Liu, Y., Zhang, Q. & Ni, M. (2010). Opportunity-based topology control in wireless sensor networks, *Parallel and Distributed Systems* pp. 405–416.

Ma, J., Chen, Q., Qian, Z. & Ni, L. (2008). Opportunistic transmission based QoS topology control in wireless sensor network, *Mob. Ad Hoc and Sen. Sys.* pp. 422–427.

Madan, R. & Lall, S. (2006). Distributed algorithms for maximum lifetime routing in wireless sensor networks, *IEEE Transactions on Wireless Communications* 5: 2185 – 2193.

Meguerdichian, S., Koushanfar, F., Potkonjak, M. & Srivastava, M. B. (2001). Coverage problems in wireless ad hoc sensor networks, *Proc. of IEEE Infocom* .

Mehrjoo, S., Aghaee, H. & Karimi, H. (2011). A novel hybrid GA-ABC based energy efficient clustering in wireless sensor network, *Multimedia and Wireless Networks* 2: 40–45.

Moscato, P. (1999). *Memetic algorithms: a short introduction*, McGraw-Hill.

Nieberg, T. (2006). *Independent and dominating sets in wireless communication graphs*, PhD thesis, Twente University, Netherlands.

Okdem, S. & Karaboga, D. (2009). Routing in WSN using an ant colony optimization (ACO) router chip, *Sensors* pp. 909–921.

Patel, M., Chandrasekaran, R. & Venkatesan, S. (2006). Energy-efficient capacity-constrained routing in wireless sensor networks, *Int. J. Perv. Comp. and Comm.* pp. 69–80.

Pelusi, L., Passarella, A. & Conti, M. (2006.). Opportunistic networking: data forwarding in disconnected mobile ad hoc networks, *Communications Magazine, IEEE* .

Rabbat, M. & Nowak, R. (2004). Distributed optimization in sensor networks, *IPSN* pp. 20–27.

Ran, G., Zhang, H. & Gong, S. (2010). Improving on leach protocol of wireless sensor networks using fuzzy logic, *Journal of Information & Computational Science* 7: 767–775.

Ren, H., Meng, M. Q. & Chen, X. (2006). Investigating network optimization approaches in wireless sensor networks, *in Proc. of IROS* pp. 2015–2021.

Rodoplu, V. & H., M. T. (1999). Minimum energy mobile wireless networks, *Journal on selected areas in communications* 17: 1333–1344.

Rossi, A., Singh, A. & Sevaux, M. (2010). Génération de colonnes dans le réseaux de capteurs sans fil, *In Proc. of ROADEF* .

Rotar, C., Risteiu, M., Ileana, I. & Hutanu, C. (2009). Optimal sensors network layout using evoutionary algorithms, *Proc. of Inter. Conf. on Automation & information* pp. 88–93.

Saleem, K., Fisal, N., Baharudin, M., Hafizah, S., Kamilah, S. & Rashid, R. (2010). Colony inspired self-optimized routing protocol based on cross layer architecture for WSN, *Int. Conf. on Communications* pp. 178–183.

Shor, N. Z. (1985). *Minimization Methods for Non-differentiable Functions*, Springer-Verlag.

Sridharan, A. & Krishnamachari, B. (2004). Max-min fair collision-free scheduling for wireless sensor networks, *Perfor., Comp. and Communic.* pp. 585–590.

Suomela, J. (2009). *Optimisation Problems in Wireless Sensor Networks: Local Algorithms and Local Graphs*, PhD thesis, University of Helsinki, Finland.

Tao, W., Chen, C., Yang, B. & Guan, X. (2010). Adaptive topology control for throughput optimization in wireless sensor networks, *Proc. of Int. Conf. Comm. Technology* pp. 1299 – 1302.

Valli, R. & Dananjayan, P. (2008). Utility enhancement by power control in WSN with different topologies using game theoretic approach, *ICCIT* 2011: 85–89.

Verdone, R. & Fabri, F. (2010). Sociability based routing for environmental opportunistic networks, *Advances in Electr. and telecom.* 1: 98–103.

Wan, P., X., L. & Wang, Y. (2001). Power efficient and sparse spanner for wireless ad hoc networks, *In IEEE Int. Conf. on Comp. Com.Net.* pp. 564 – 567.

Wang, Q., Fan, P., Wu, D. & Ben Letaief, K. (2011). End-to-end delay constrained routing and scheduling for wireless sensor networks, *ICC* pp. 1–5.

Wang, T., Wu, Z. & Mao, J. (2007). A new method for multi-objective TDMA scheduling in WSN using pareto-based PSO and fuzzy comprehensive judgement, *Proc. of High Perf. Comp. and Comm.* pp. 144–155.

Wang, Y., Wang, W., Li, X. & Song, W. (2008). Interference-aware joint routing and TDMA link scheduling for static wireless networks, *IEEE Trans. on par. and distrib. sys.* 19: 1709–1725.

Wang, Z. & Crowcroft, J. (1996). Quality of service routing for supporting multimedia applications,, *IEEE J. Sel. Areas Commun* 14: 1228–1234.

Wu, Y. & Li, Y. (2008). Construction algorithms for k-connected m-dominating sets in WSN, *MobiHoc* pp. 83–90.

Wu, Y., Stankovic, J. A., He, T., Lu, J. & Lin, S. (2008). Realistic and efficient multi-channel communications in wireless sensor networks, *INFOCOM* pp. 1193–1201.

Xu, W., Chen, J., Zhang, Y., Xiao, Y. & Sun, Y. (2008). Optimal rate routing in wireless sensor networks with guaranteed lifetime, *IEEE Globcom* pp. 1–5.

Xue, Y., Cui, Y. & Nahrstedt, K. (2005). Maximizing lifetime for data aggregation in wireless sensor networks, *Mobile Networks and Applications* 6: 853 – 864.

Yang, H. & Cai, W. (2008). Distributed power control algorithm with multi-QoS constraints for wireless sensor networks, *Proc. of Int. Conf on Netw., Sens. and Contr.* pp. 1031–1036.

Ye, W. & Heidemann, J. (2003). Medium access control in wireless sensor networks, *Technical report*, ISI-TR-580, Information Sciences Institute.

Yilmaz, O. & Erciyes, K. (2010). Distributed weighted node shortest path routing for wireless sensor networks, *Communications in Computer and Information Science*, 84: 304–314.

Yoneki, E., Hui, P., Chan, S. & Crowcroft, J. (2007).   A socio-aware overlay for publish/subscribe communication in delay tolerant networks, *Proc. of MSWiM, ACM* pp. 225–234.

Yu, Q., Chenyz, J., Fanz, Y., Shenz, X. & Suny, Y. (2010). Multi-channel assignment in wireless sensor networks: A game theoretic approach, *INFOCOM* pp. 1127–1135.

Yuanyuan, Z., Jia, X. & Yanxiang, H. (2006). Energy efficient distributed connected dominating sets construction in WSN., *Proc. of Wireless communications* pp. 797–802.

Zou, Y. & Chakrabarty, K. A. (2005).   A distributed coverage and connectivity centric technique for selecting active nodes in wireless sensor networks., *IEEE Trans. on Comp.* 54: 978–991.

Permissions

The contributors of this book come from diverse backgrounds, making this book a truly international effort. This book will bring forth new frontiers with its revolutionizing research information and detailed analysis of the nascent developments around the world.

We would like to thank Professor Jesús Hamilton Ortiz, for lending his expertise to make the book truly unique. He has played a crucial role in the development of this book. Without his invaluable contribution this book wouldn't have been possible. He has made vital efforts to compile up to date information on the varied aspects of this subject to make this book a valuable addition to the collection of many professionals and students.

This book was conceptualized with the vision of imparting up-to-date information and advanced data in this field. To ensure the same, a matchless editorial board was set up. Every individual on the board went through rigorous rounds of assessment to prove their worth. After which they invested a large part of their time researching and compiling the most relevant data for our readers. Conferences and sessions were held from time to time between the editorial board and the contributing authors to present the data in the most comprehensible form. The editorial team has worked tirelessly to provide valuable and valid information to help people across the globe.

Every chapter published in this book has been scrutinized by our experts. Their significance has been extensively debated. The topics covered herein carry significant findings which will fuel the growth of the discipline. They may even be implemented as practical applications or may be referred to as a beginning point for another development. Chapters in this book were first published by InTech; hereby published with permission under the Creative Commons Attribution License or equivalent.

The editorial board has been involved in producing this book since its inception. They have spent rigorous hours researching and exploring the diverse topics which have resulted in the successful publishing of this book. They have passed on their knowledge of decades through this book. To expedite this challenging task, the publisher supported the team at every step. A small team of assistant editors was also appointed to further simplify the editing procedure and attain best results for the readers.

Our editorial team has been hand-picked from every corner of the world. Their multi-ethnicity adds dynamic inputs to the discussions which result in innovative outcomes. These outcomes are then further discussed with the researchers and contributors who give their valuable feedback and opinion regarding the same. The feedback is then collaborated with the researches and they are edited in a comprehensive manner to aid the understanding of the subject.

Apart from the editorial board, the designing team has also invested a significant amount of their time in understanding the subject and creating the most relevant covers. They scrutinized every image to scout for the most suitable representation of the subject and create an appropriate cover for the book.

The publishing team has been involved in this book since its early stages. They were actively engaged in every process, be it collecting the data, connecting with the contributors or procuring relevant information. The team has been an ardent support to the editorial, designing and production team. Their endless efforts to recruit the best for this project, has resulted in the accomplishment of this book. They are a veteran in the field of academics and their pool of knowledge is as vast as their experience in printing. Their expertise and guidance has proved useful at every step. Their uncompromising quality standards have made this book an exceptional effort. Their encouragement from time to time has been an inspiration for everyone.

The publisher and the editorial board hope that this book will prove to be a valuable piece of knowledge for researchers, students, practitioners and scholars across the globe.

List of Contributors

F. Pereniguez-Garcia, R. Marin-Lopez and A.F. Gomez-Skarmeta
Faculty of Computer Science, University of Murcia, Spain

M. Bellafkih, B. Raouyane and M. Errais
Institut National des Postes et Télécommunications, Rabat, Morroco

B. Raouyane, M. Errais and M. Ramdani
Faculté des Sciences et Techniques, Mohammedia, Morroco

D. Ranc
IT Sud Paris, Evry, France

Paulo H. P. de Carvalho, Márcio A. de Deus and Priscila S. Barreto
Departament of Electrical Engineering, Departament of Computer Science, University of Brasilia, Brazil

A. Toppan, P. Toppan, C. De Castro and O. Andrisano
IEIIT-CNR, National Research Council of Italy & WiLab, University of Bologna, Bologna, Italy

Wei Zhuang
China Telecom Co. Ltd. (Shanghai), P.R.China

C. Guerra Torres and J. de León Morales
Facultad de Ingenieria Mecánica y Eléctrica, Universidad Autónoma de Nuevo León, México

Khadija Stewart
DePauw Universtity, USA

James L. Stewart
Purdue University, USA

Ada Gogu and Dritan Nace
Université de Technologie de Compiègne, France

Arta Dilo and Nirvana Meratnia
University of Twente, The Netherlands